NANOBIOTECHNOLOGY

MJP
PUBLISHERS

NANOBIOTECHNOLOGY

Subbiah Balaji

Lecturer
PG and Research Department of Microbiology
Hindusthan College of Arts and Science
Coimbatore, TN

MJP PUBLISHERS

Chennai Trichy Tirunelveli NewDelhi

MJP PUBLISHERS

ISBN 978-81-8094-020-0
MJP PUBLISHERS
No.44, Nallathambi Street
Triplicane, Chennai 600 005

Printed and bound in India
MJP 069

Publisher : C. Janarthanan

To the loving memory of my father, Mr. S. Subbiah

and

to my mother, Mrs. S. Kothai Subbiah and brother, Subbiah Kannapiran

FOREWORD

The Science and Technology of the materials at the nanometre scale has emerged as one of the most promising subjects in the past two decades. One nanometre is one billionth of a metre; materials ranging down to nanometre scale exhibit properties completely different from their bulk counterparts. The exotic properties of the nanomaterials enable their applications in various fields such as high-speed electronic devices, sensors, bio-diagnostic devices, energy conversion and storage, bio-compatible materials and devices and also for environmental remediation. It has been demonstrated that the properties of the nanoscale materials can be tailored and further the nanomaterials can be integrated together to form a functional device. Nanoscience and Technology is a multidisciplinary field comprised of physics, chemistry, biology and engineering. Therefore, basic knowledge of physics, chemistry and biology is needed to establish an understanding of the phenomenon taking place at the nanometre range. Such a vast and multidisciplinary field needs a broad overview of the role of nanotechnology in future emerging applications.

This book provides a broad overview of bionanotechnology, starting from basics to various applications such as biomaterials, nanomedicine, drug delivery and bio-medical devices. Besides the merits of bionanotechnology, the author has presented possible envi0ronmental threats due to nanoparticle contamination.

I am sure this book will be useful to the common man and the beginners who want to gain a broad overview in bionanotechnology. This book is suitably structured that the common man can understand and appreciate the latest developments in bionanotechnology.

Prof. Dr. G. Thiruvasagam
Vice-Chancellor
University of Madras

PREFACE

During the last few years the terms nanotechnology and nanobiotechnology have become increasingly popular and it appears that these fields would be deciding the fate of the 21st century.

This book is designed specifically to provide an overview of nanosized biological materials and substances for today's graduate students and provides the current information on different aspects of nanobiotechnology.

This book is an outcome of the compilation of specific research work done by various eminent scientists and researchers in different areas of nanobiotechnology from reputed research laboratories across the world.

The contents of the book have been organised in six chapters.

Chapter 1 gives a basic introduction to nanotechnology and discusses its background in detail. It also discusses the market strategies of nanotechnology and future prospects of nano- manufacturers.

Chapter 2 discusses the basics of nanomaterials. Biodegradable polymers and their role in biological recognition and tissue engineering are discussed in detail in this chapter.

Chapter 3 deals with the construction/fabrication and characterization of nanomaterials. The synthesis methodology of nanoparticles and the various techniques which help to evaluate the synthesized nanomaterials have been discussed.

Chapter 4 provides basic information of micro- and nanoscale devices in biomedical applications. Biomedical sensors and biosensors, and quantum dot technology are also discussed in detail to enhance one's knowledge of drug delivery systems for cancer therapy.

Chapter 5 primarily deals with the concept of nanomedicine and novel drug delivery systems which are specifically connected to cancer dignosis methods. Various types of delivery vehicles such as microcapsules and microspheres which are in the micro-range in size and micelles, liposomes, hydrogels and dendrimers which are in the nano-range have also been discussed.

Chapter 6 focuses on the impact of nanotechnology, and discusses the adverse health effects that may be caused due to nanoparticle exposure while handling materials during physical and chemical processes of nanomaterial synthesis, which may lead to discharge of unwanted nano by-products. The solution to overcome this problem by using microbes and plants as nanofactories has also been discussed in detail.

All chapters have a "Concept Check" and "Mind Twister" section at the end to help readers to evaluate their understanding of the subject.

Finally, I personally take responsibility for any kind of errors that may be present in this book. I look forward to valuable comments, suggestions and constructive criticisms from the readers, which may be sent to me at balaji_bio@rediffmail.com.

Subbiah Balaji

ACKNOWLEDGEMENTS

Fisrt and foremost I thank the Almighty God for his endless blessings and for providing a direction in life that has made it possible for me to reach this far.

Words cannot adequately express my admiration and heartfelt gratitude to all those who helped me in bringing out this book. It is a pleasure to thank Thiru. Khannaiyan, Chairman; Sarasuwathi Kannaiyann, Secretary, and Priya Satish Kumar, Joint Secretary of Hindusthan Group of Institutions and Dr. N. Baluswami, Principal, Hindusthan College of Arts and Science, for their constant support and encouragement.

I would like to express my sincere thanks to Dr. T. Rajasekaran, Researcher and Lab Head, Department of Biochemistry, D1 Oils Limited, Praia, Capeverde, West Africa. His valuable suggestions gave so many ideas to finish my work successfully.

It is my duty to thank Dr. N. Murugalatha, Head, PG and Research Departments of Biotechnology and Microbiology, and all the teaching staff for their critical evaluation of my writing and their innovative ideas which helped me complete my work in a systematic way.

I am deeply grateful to the wonderful people at MJP Publishers, including Mr. J.C. Pillai, Director, for accepting to publish this book and Mr. C. Sajeesh Kumar, Managing Editor, for his valuable guidance and moral support.

Last but not least, I have no words to express my gratitude to my mother who loves me more than anyone in the world. My mother's love and my wife Revathi's support have always been my strength. I thank them for showering their undemanding love on me.

Subbiah Balaji

CONTENTS

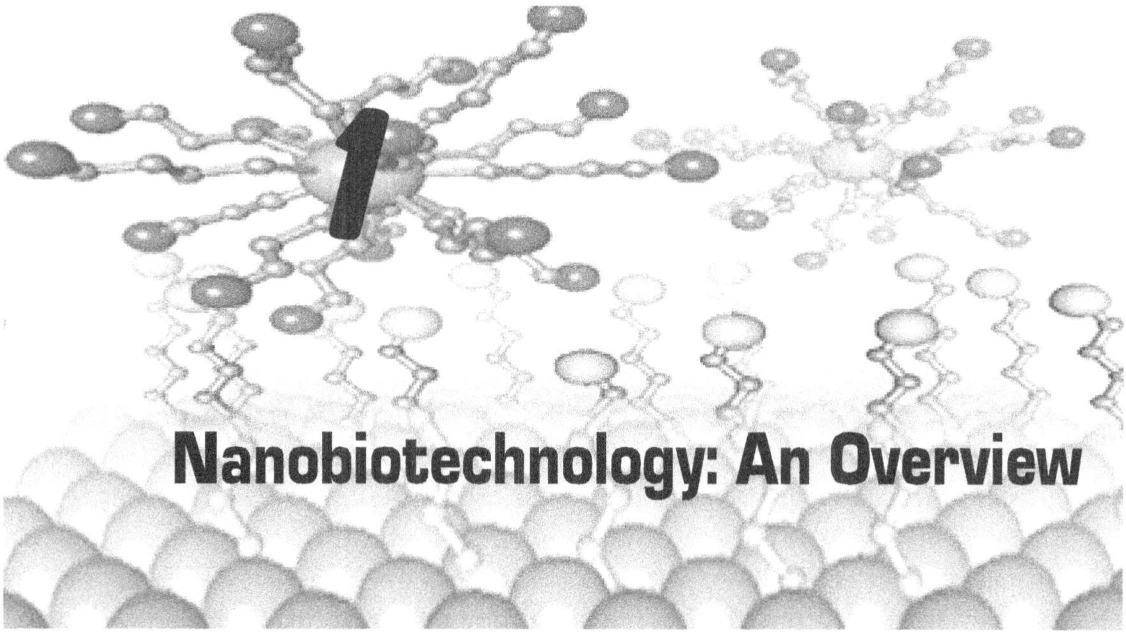

1 Nanobiotechnology: An Overview

WHAT IS NANOTECHNOLOGY?

Nano is a Greek prefix which means dwarf. In science this prefix denotes a fraction of 10^{-9} a given unit. For instance 1 nm = 10^{-9} m. **Technology** means the discipline dealing with the art or science of applying scientific knowledge to practical problems. **Science** is the knowledge or a system of knowledge concerned with the physical world and its phenomena.

Nano + Technology = Nanotechnology

Nano + Science = Nanoscience

Nanoscience is the actual "science" or basic study of systems and materials at the nanoscale, whereas, **nanotechnology** is the application of nanoscience to a broad set of emerging manufacturing technologies, which control and manipulate material at the level of atoms and molecules. In practice, the term Nanotechnology is used in a broad sense and includes Nanoscience.

This chapter summarizes the key findings of Nanotechnology and the great potential benefits of nanotechnologies, which are attracting and increasing investment from governments and industries around the world. Total global expenditure is approximately around \$8.75 billion at present, but this is expected to increase. The USA's 21st Century Nanotechnology Research and

Development Act (2003) allocated almost $3.7 billion to fund nanotechnologies during 2005–2008 as compared to just $750 million spent in 2003. Between 2001 and 2003, the Japanese Government doubled its nanotechnology funding to $800 million. Within Europe, about $1.25 billion per annum is currently spent on nanotechnology research and development, and the UK Government has allocated about $81.9 million per year from 2003 to 2009. Analysts have estimated that the size of the market will be $13 billion in 2010.

BACKGROUND

Nanoscience may be defined as the study of phenomena and the manipulation of materials at the atomic, molecular, and macromolecular scales, here properties differ significantly at larger scale, and nanostructured materials are always based on design, characterization, production, and application of structures, devices, and systems by controlling shape and size at the nanometre scale. These definitions involve many traditional scientific disciplines and encompass a broad and varied range of materials, tools, and approaches. It can be said that the only feature common to the diverse activities that have been characterized as "nanotechnology" is the tiny dimensions on which they operate. In nature too, there are structures and processes operating at the nanoscale, and range from simple colloids such as milk to complicated proteins. Combustion and cooking also release free nanoparticles as by-products. In some sense, nanoscience and nanotechnologies are not new: size-dependent properties have been exploited for centuries. For example, Au and Ag nanoparticles (particles of diameter less than 100 nm) have been used as coloured pigments in stained glass and ceramics since the 10th century AD. In fact, nanoparticles were used over 2000 years ago in Roman glass, where clusters of Au nanoparticles were used to generate vivid colours.

Many chemicals and chemical processes have nanoscale features, and, for example, chemists have been making polymers—large molecules made up of nanoscalar subunits—for many decades. Nanotechnologies have been used to create the features on computer chips for the past 20 years. However, with the invention of imaging techniques like the scanning tunnelling microscope and the atomic force microscope, our understanding of the nanoworld has increased multifold and it is now becoming possible to control structures at the nanoscale.

MILESTONES IN NANOTECHNOLOGY

Year	Scientist	Discovery
3.5 Mrd. Years		First cells with nano machines
400 BC	Demokrit	Reasoning about atoms and matter
1905	Albert Einstein	Calculation of molecular diameter
1931	Max Knoll and Ernst Ruska	Electron microscope
1959	Richard Feynman	There's Plenty of Room at the Bottom
1968	Alfred Y. Cho and John Arthur (Bell Labs)	MBE (atomic layer growth)
1974	Norio Taniguchi	Nanotechnology for fabrication methods below 1 μm.
1981	Gerd Binning and Heinrich Rohrer	Nobel Prize for scanning tunnelling microscope
1985	Robert F. Curl, Harold W. Kroto and Richard Smalley	Buckminsterfullerenes (Buckyballs)
1986	Eric K. Drexler	Engines of Creation
1989	Eigler M.	Writing with an STM tool
1991	SumioIigima (NEC)	Carbon nanotubes
1993	Warren Robinett and Stanley R. Williams	Combination of SEM and VR (virtual reality system)
1998	Cees Dekker *et al.*	Carbon nanotube transistor
1999	James M. Tour and Mark A. Read	Single molecule switch
2000	Eigler *et al.*	Construction of Quantum Corrals and Quantum mirrors
2001	Florian Bamberg	Soldering of nanotubes with e-beam
2004	Intel Corporation	Launching of Pentium IV "Prescott" processor based on 90 nm technology

OPPORTUNITIES

A large number of potential applications of nanotechnologies are now opening up. In this chapter there is an overview of current and potential future developments in nanoscience and nanotechnologies. Much of nanoscience and many nanotechnologies are concerned with producing new or enhanced materials. Some nanotechnology-enabled products are already on the market and are commercially very successful. For example, self-cleaning windows use a 15 nm thick coating of activated TiO_2 engineered to be highly water-repellent, so that rainwater just flows off the surface, washing away the dirt. Nanoparticles are used in some sunscreens to reflect and absorb ultraviolet (UV) light. Carbon nanostructures have been the focus of much interest and research since they were first observed in the mid-1980s. The football-shaped buckminsterfullerene (C_{60}) and its analogs show great promise as lubricants and, they have cage structures, which makes them useful as drug delivery systems, as well as in electronics. The same graphite sheet structure, which allows electrical conductivity, was discovered in carbon nanotubes in the early 1990s.

Carbon nanotubes are only a few nanometres in diameter, but they can be made up to several centimetres long, and this gives them unique properties—they are as stiff as diamond and also conduct electricity. Many applications of carbon nanotubes in reinforced composites, sensors, nanoelectronics, and display devices are being conceived. Around 100 tonnes are currently being manufactured per year, but it is still difficult to guarantee specific dimensions and physical properties and to separate the tubes from the tangles in which they emerge from the production process. It is also possible to make similar structures from inorganic substances such as MoS_2 and TiO_2, which may be useful as catalysts and for energy storage. Future applications of nanomaterials include lighter, stronger materials, the use of nanoparticles to clean up contaminated land, and nanoengineered membranes for more energy-efficient water purification or desalination.

Computer chips and CD and DVD drives are already operating at nanoscales, and nanoscience and nanotechnologies will continue to play a crucial role in the progressive miniaturization of computer chips and the enhancement of data storage. It could also enable advances such as roll-up TV screens. Nanotechnologies are also enabling the development of smaller, cheaper sensors, which will have a wide range of applications from monitoring the pollution in the environment, the freshness of food, or the stresses in a building or a vehicle. Much interest is also focused on quantum dots, which are semiconductor nanoparticles that can be "tuned" to emit or absorb particular colours of light for use in solar energy or fluorescent biological labels.

Applications of nanotechnologies in medicine are especially promising in the longer term. These can be expected to enable drug delivery targeted at specific sites in the body so that, for example, chemotherapy is less invasive. Nanotechnology is expected to lead to stronger, longer-lasting implants; sensors that can be used to monitor aspects of human health; and improved artificial cochleae and retinas. However, many of these applications will not be realized for at least ten years, partly because of the rigorous testing and evaluation that will be required. Antimicrobial wound dressings are already on the market in the USA. These use nanocrystalline Ag to provide a steady dose of ionic Ag to protect against secondary infections and are claimed to be effective against 150 different pathogens. An interesting aspect of the current decade for nanotechnologies is that, for the first time, there is a convergence between the two approaches to creating nanostructures, particles, and systems. Top-down manufacturing, perfected by the semiconductor industry over the last 30 years, involves precision engineering and lithography to grind or cut bulk materials into tiny pieces and to etch or print nanoscale patterns onto them. Bottom-up manufacturing assembles nanostructures from scratch through conventional chemical synthesis, self-assembly (harnessing natural physical or chemical bonding to assemble structures like crystal formation), and positional assembly (where atoms or groups are manipulated individually into a structure). The two approaches can now control product dimensions of a similar order, which opens up exciting new possibilities in hybrid manufacturing.

However, scaling-up production from the laboratory to mass manufacturing is not easy. In the longer term, it is hoped that nanotechnologies will enable more efficient approaches to manufacturing using less raw materials and energy. It is important to substantiate such claims, so a series of life-cycle assessments should be undertaken for the applications and product groups arising from nanotechnologies to ensure that what is saved in one part of the life cycle is not offset by increases in another.

NANOPARTICLES AS BUILDING BLOCKS

Nanomaterials and, in particular, nanoparticles are at the forefront of the nanotechnology wave. Over recent years there has been a rapid increase in the range of available nanomaterials and the number of companies that supply them. A variety of production techniques is used, each with benefits and drawbacks, but these are being improved all the time. The current and potential applications for nanoparticles are growing and cover an extremely broad range of markets. This is causing a great deal of interest among companies, but there are still a number of hurdles to be overcome before the potential of nanoparticles is fully realized.

But nanotechnology is not new and, although the term was first coined in the 1960s, it can be argued that material scientists and chemists have been working in nanotechnology since their disciplines started. The real move towards the use of nanoparticles did not occur until the early 20th century with the production of carbon black and, subsequently, fumed silica in the 1940s. However, it has only been with the development of high-speed computing, and hence modelling; advanced characterization techniques, such as atomic force microscopy and scanning tunnelling microscopy; and synthesis routes, such as sol-gel processing, that it has been possible to design nanomaterials. With these materials have come significant improvements in the performance of polymeric materials. However, this was achieved through an empirical approach—the materials were not designed but discovered. It is this differentiation that distinguishes modern nanotechnology from previous activities in materials science and chemistry. As a recent European Union report stated, "Nanotechnology is expected to create new materials with totally new, problem-solving properties."

WHAT ARE NANOPARTICLES?

Although nanotechnology is widely talked about, there is little consensus about where the nano domain begins. In a recent report published in Economist Intelligence Unit and the Institute of Nanotechnology, around one hundred nanotechnology experts were surveyed and the results of the survey were as shown in Table 1.1.

TABLE 1.1 VARIOUS VIEWS OF NANOTECHNOLOGY

Definitions of nanotechnology given by experts in a survey	Percentage of respondents
Technology that works with elements up to 100 nm in size	45
Technology that operates at the level of atoms and molecules	23
Technology that works with elements on a sub-micron level	17
Technology that operates under new laws of physics	5
Other responses	10

WHY ARE NANOPARTICLES INTERESTING?

Just because these materials can be made into very small particles, it does not mean that they have any immediate practical use. However, the fact that these materials can be made at this minute scale gives them the potential to exhibit some very interesting properties. Materials with particle sizes on the nanoscale between 1 nm and 250 nm lie in the domain between the quantum effects of atoms and molecules and the bulk properties of materials. Therefore, physical properties of materials are controlled by phenomena that have critical dimensions on the nanoscale. The ability to fabricate and control the structure of nanoparticles allows the scientist and engineer to influence the resulting properties and, ultimately, design materials to give the desired properties. There is an extremely wide range of applications where the physical size of the particle can provide enhanced properties that are of benefit, for example:

- the small size allows finer polishing and smoother surfaces;
- where the grain size is too small for dislocations, high-strength, high-hardness metals can be made; and
- the high surface area enables the production of more efficient catalysts and more energetic materials.

HOW ARE THEY MADE?

The manufacture of nanoparticles can be achieved through different routes. Some of these have been in existence for many years; others are far more recent. There are four generic routes to make nanoparticles: wet chemical; mechanical; form-in-place; and gas phase synthesis. Each of these routes need to be studied, as the resultant materials can have significantly different properties, depending on the fabrication route that is chosen. Also, some routes are more suited to the fabrication of certain classes of materials than others.

Wet Chemical Processes

These include colloidal chemistry, hydrothermal methods, sol-gels, and other precipitation processes. Essentially, solutions of different ions are mixed in well-defined quantities and under controlled conditions of heat, temperature, and pressure to promote the formation of insoluble compounds that precipitate out of the solution. These precipitates are then collected through filtering and/or spray-drying to produce a dry powder. The advantages of these wet chemical processes include the following.

- A large variety of compounds can be fabricated, including inorganics and organics, as well as some metals, in essentially cheap equipment and significant quantities.
- It is possible to control particle size closely and to produce highly monodisperse materials.

However, there are limitations on the range of compounds that are feasible: bound water molecules can be a problem and, for sol-gel processing in particular, yields can be quite low. So, for bulk production, large quantities of starting materials may be required, which could be expensive.

Mechanical Processes

These include grinding, milling, and mechanical alloying techniques. These processes use the age-old technique of physically pounding coarse powders into finer and finer powders, similar to flour mills. Today, the most common processes are either planetary or rotating ball mills. The advantages of these techniques include the following.

- they are simple,
- they require low-cost equipment, and
- many materials are capable of being processed provided that a coarse feedstock powder can be made.

Some difficulties encountered are:

- agglomeration of the powders, which leads to broad particle size distributions,
- contamination from the process equipment itself, and
- the achievement of very fine particle sizes.

Commonly, these methods are used for inorganic and metals, but not organic materials.

Form-in-Place Processes

These include lithography, vacuum deposition (physical vapour deposition and chemical vapour deposition), and spray coatings. These processes are used more in the production of nanostructured layers and coatings, but can be used to fabricate nanoparticles by scraping the deposits from the collector. However, they tend to be quite inefficient and are generally not used for the fabrication of dry powders, although some companies are beginning to exploit these processes.

Gas Phase Synthesis

This includes flame pyrolysis, electroexplosion, laser ablation, high-temperature evaporation, and plasma synthesis techniques. Flame pyrolysis has been used for many years in the fabrication of simple materials such as carbon black and fumed silica, and is being used in the fabrication of many other compounds. Laser ablation is capable of making almost any nanomaterial, since it utilizes a mix of physical erosion and evaporation, but the production rates are extremely slow and most suited to research use. Both RF (radio frequency) and DC (direct current) plasmas are being used successfully to make a wide range of materials. The heat source is very clean and controllable, and the temperature in the plasmas can reach 9000°C or even more, which means that even highly refractory materials can be processed. But, this also means that these methods are not suitable for processing organic materials.

Thus different methods can be employed to manufacture nanoparticles. All of them are being used commercially and each has its own merits and drawbacks. However, we may not be able to say which methods will dominate, since it is the demand for specific materials that will drive the adoption of a particular manufacturing process.

THE MARKET FOR NANOPARTICLES

A common technique for changing the properties of materials is by adding particulate materials to a matrix. This has been in practice almost since the first synthetic materials were developed. However, the size of the additives was usually larger than the nanoscale. The first industrial production of a nanomaterial was the production of carbon black early in the 20th century. Subsequently, in the 1940s fumed silica was produced. These materials are still produced and used in vast quantities, and some well-known companies such as Degussa and Cabot base their business on these materials. But it was only in the latter part of the 20th century that the scientific understanding of materials incorporating ultrafine particulates really developed and it was realized that significant improvements to properties could be achieved.

The real burst in the commercialization of nanoparticle production has occurred only over the last ten years or so. Where the data are available, one can see this rapid growth (Figure 1.1). The number of companies doubled in the 1990s,

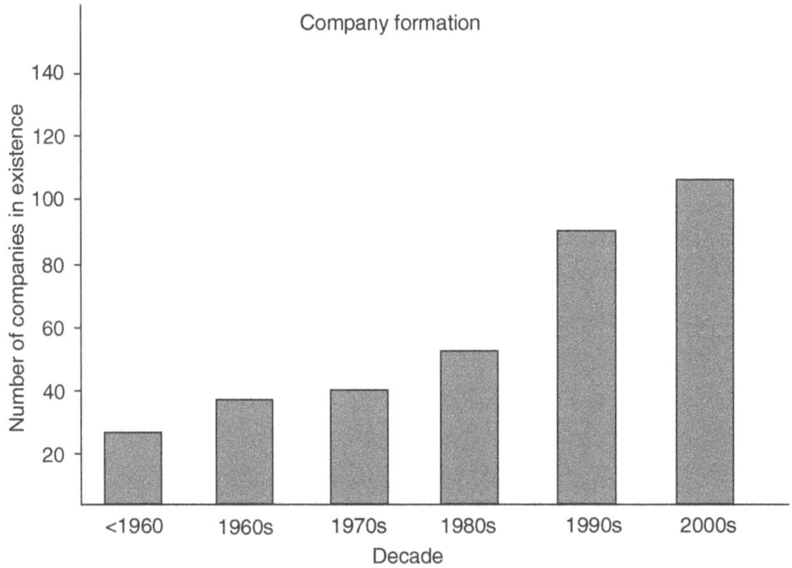

FIGURE 1.1 GROWTH IN NUMBER OF NANOMATERIAL COMPANIES

and the current rate of formation appears to maintain the same pace if not exceed this in the coming years. The establishment of new, small start-up companies that have been spun-out of universities, government laboratories, or set up by entrepreneurs have facilitated this recent growth and, it is now estimated that over half of all nanomaterial companies fall into this category. Although accurate data is difficult to obtain, there are probably more than 420 companies producing nanomaterials in various forms in the world today, of which about 200 are nanoparticle producers. The split between types of material is shown in Table 1.2.

However, not all these companies sell the nanomaterials that they make and many are geared up to generate and license intellectual property based on the use of their nanomaterials. In addition to these companies, there are others that make nanomaterials for use in their own products.

As the benefits of nanomaterials are realized, there will be a greater progression in the evolution of nanoscale materials. However, the growth in interest cannot be explained entirely by such evolutionary development. As more research effort is directed towards investigating nanoscale materials, the ability

TABLE 1.2 NANOMATERIAL PRODUCT TYPES AND MARKET FOCUS OF COMPANIES

Types of products	Number
Nanoparticles	160
Nanotubes	55
Nanoporous materials	22
Fullerenes	21
Quantum dots	19
Nanostructured materials	16
Nanofibres	9
Nanocapsules	8
Nanowires	6
Dendrimers	5
Total	321

to make a step-change in performance is being found all the time. This is opening up totally new applications and the possibility of making products that have been hypothesized for many years, such as targeted drug delivery methods, new optoelectronic devices, and smaller, more efficient energy devices. This has given rise to some widely differing estimates of the size of the nanotechnology market and the nanomaterials subset of this.

FUTURE FOR NANOPARTICLES AND THEIR MANUFACTURERS

There is going to be a sure escalation in the use of nanoparticles and the market for nanoparticles has the potential to increase dramatically over the next ten years as more uses for these materials are developed and commercialized. A major impact will be in the medical and pharmaceutical markets as new treatments relying on nanoparticles obtain licenses for use. However, there are many other applications, like for example, consumer goods, where the time taken for the product to reach the market is much less than for the pharmaceutical and medical products. But the nanomaterials companies are still faced by many challenges and they need to overcome these so that their full potential can be realized. These challenges include the following.

- *How to produce materials at a large scale at feasible prices*—many current techniques cannot scale-up sufficiently to produce the cost reductions required to target volume markets.
- *How to supply the materials in a form suitable for inclusion in manufacturing processes*—understanding the surface chemistry and how particles can be dispersed in a wide variety of media will be key to the adoption of many materials.
- *Consistency and reliability in large-scale production*—tolerances on size and composition can be achieved reliably for simple compounds such as binary oxides and for more complex materials in small batch production, but doing this for complex materials in large-scale manufacture is not so easy.
- *Characterization*—it is possible to characterize materials to a great extent. However, many of the techniques are appropriate for the research laboratory, but not for the production environment. What is required is rapid, bulk, and, preferably, on-line techniques to monitor properties such as particle size distribution.
- The need to focus in a very broad market will determine the survival of many nanomaterials companies in the short term as they start to build revenues.
- Adding and retaining value will be key to the longer-term viability of companies as volumes and the pressure to reduce prices and margins increase. The approach being adopted by many is that of securing intellectual property to provide a longer-term income stream.
- *Health, safety, and environment*—the profile of nanotechnology has increased in recent months, with a focus on the potential long-term effects of nanotechnology and, more immediately, of nanomaterials on humans and the environment. As with any high-profile technology, questions will be asked, but some nanomaterials have been with us for many years without causing concerns. However, it is very important to the success of this industry that any concerns are addressed. The key aspect is: are there any detrimental effects, over and above those already identified for the materials, which occur purely from the fact that these materials are in the nano-form? It is also highly unlikely that nanomaterials will be used without being incorporated into some other media, such as a composite or liquid. Research is under way into the effects of nanomaterials and it is difficult to draw any firm conclusions to date, but there is evidence that there may be positive benefits from these types of materials, both for humans and the environment.

Nanotechnology as a whole is estimated to represent a market of $11 trillion by 2010, with nanomaterials growing from $490 million to $900 million in 2005 and $11 billion in 2010. However, the impact of nanomaterials will extend well beyond the immediate value of the materials themselves. One estimate has been made that nanostructured materials and processes could have an impact of over $340 billion by 2010.

BIOINSPIRED APPROACHES TO BUILDING NANOSCALE DEVICES

In 1959, Richard Feynman gave a speech called "There's plenty of room at the bottom," where he challenged scientists to construct atomic-scale devices. Since 1959, the dream of building nanoscale devices ($\sim 1 \times 10^{-9}$ m) is slowly becoming a reality. The expected impact of nanodevices is immense. Nanodevices are comparable in size to most biological systems (e.g. viruses, organelles, proteins, DNA) and therefore, it has been suggested that nanodevices can be engineered to diagnose and treat malfunctions in cells. This can lead to a new generation of treatment strategies for cancer, AIDS, or Alzheimer's disease. Future advances in the computer chip industry may also depend on nanotechnology—the ability to design atomic-scale chips may provide hard drives with greater memory capabilities or faster electronics. Assembled nanosystems are also expected to impact the aerospace program—where lightweight, high tear-resistant fabric (made with nanostructures) with embedded nanosensors may be designed for astronauts. To meet the challenges and rewards of nanotechnology, research institutions around the world have built infrastructures to study, manipulate, and design nanometre-sized materials (generally, < 100 nm) for building atomic-scale devices.

In the last couple of centuries, we have perfected the art of building macroscale devices such as clocks, calculators, computer chips, and cars. The building process generally involves the manufacturing of precursor components and the assembling of these components into a functional unit. The entire process can be automated using manmade machines. However, there is extreme difficulty in building similar devices with nanometre-dimensions. This is mainly due to the inability to assemble atoms or nanometre-sized components in a coordinated fashion. Top-down approaches, generally associated with photolithography and etching techniques, have been instrumental in advances in microscale technologies. However, a top-down approach has found limited use in building nanoscale devices mainly due to high costs of instruments and diffraction limit. A bottom-up strategy, where atoms are precisely assembled in the molecular scale, is one of the most promising strategies for building atomic-scale devices, as illustrated in Figure 1.2.

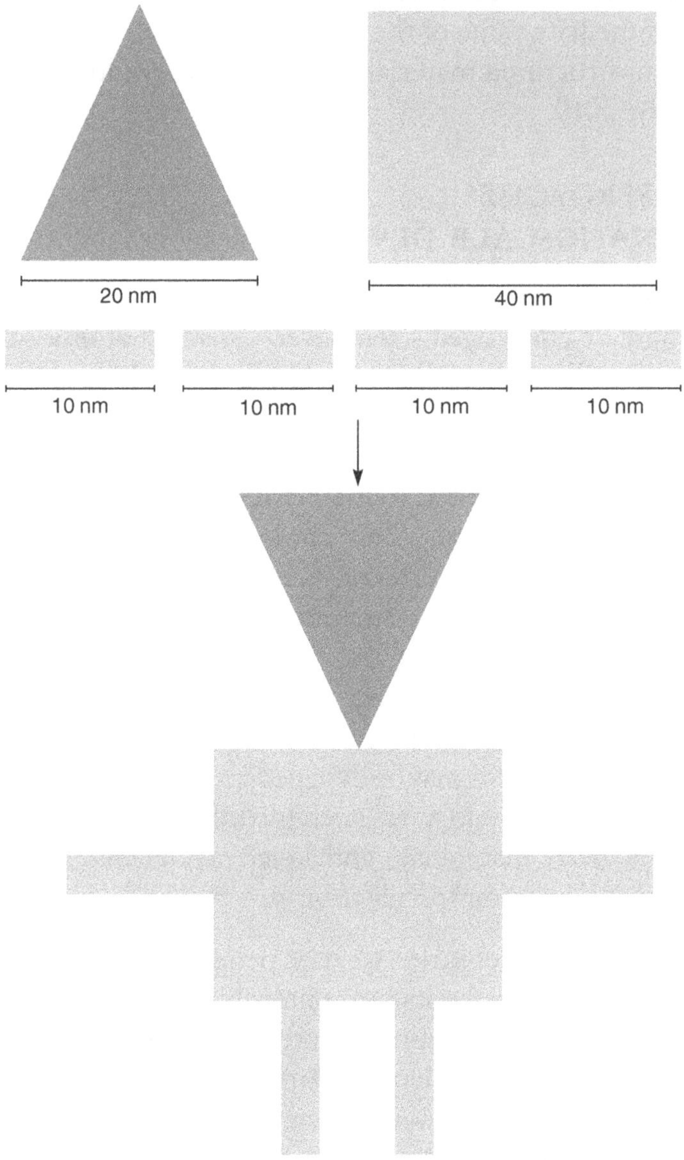

FIGURE 1.2 ASSEMBLING OF NANOSTRUCTURE

Material scientists and chemists have spent the last thirty years perfecting the synthesis of nanoscale objects, which may one day become precursors for building nanoscale devices. Semiconductor, metallic, non-metallic, and alloyed nanoparticles have been successfully synthesized and characterized. The precision of synthesis is so great that researchers can selectively and reproducibly grow protrusions on the surface of spherical quantum dots or modify the dimensions of metallic nanoparticles using solution-based approaches. Simple monofunctional hybrid organic/inorganic nanostructures have demonstrated a plethora of applications in biomedical detection. Yet, the current challenge of nanotechnologists is to mix-and-match different types of nanostructures in a coordinated fashion to produce a functional device the size of a virus (<100 nm).

Integrating organic molecules with inorganic nanostructures have produced exciting results in the field of nanodevice building. Organic molecules have been utilized as surface coatings to prevent unwanted nanoparticle aggregation, as molecules to direct nanoparticle assembly, and as homing devices to target nanostructures to specific biological sites. Furthermore, organic molecules have embarked greater functional capabilities into inorganic nanostructures. For example, oligonucleotides used to assemble and de-assemble metallic nanostructures. At the current state of research in this field, there is a wide-array of precursor nanostructures and organic molecules available to build nanodevices. However, the key challenge is to develop novel approaches to assemble them into a functional unit.

BIOLOGY AS MODEL SYSTEM FOR BUILDING NANOSCALE DEVICES

A key challenge in building nanoscale devices is the ability to assemble components in a controlled fashion. How can one build a nanometer-sized RC-circuit? How can one build a detection and drug storage and delivery system the size of a standard virus (~50 to 150 nm)? Researchers have recently looked toward biology as a guide to assemble nanostructures into functional devices. One interesting fact about bioassembly in living cells and tissues is that only a small amount of subunits are required to produce a rich and diverse group of functional systems that regulate and maintain the viability of cells, tissues, organs, and the organism. The information in DNA, the blueprint of a cell, is encoded using only four distinct bases. Proteins, the functional units of a cell, are only composed of 20 amino acids. Yet, a cell has developed the capability to assemble 4 bases into a complete genetic code and 20 different amino acids into thousands of functional proteins.

The molecular inner-workings of a cell can be equated to a highly efficient assembly line that produces many types of biological nanomachines. For example, in the translation of ribonucleic acid (RNA) to proteins, individual protein units are assembled together into a ~30-nm sized functional system called a ribosome. The ribosome interacts with RNA and translates the RNA sequence into an amino acid sequence. This amino acid sequence then interacts with other biological systems to fold into an active protein. Eukaryotic ribosome contain up to 82 proteins that assemble in a precise fashion to form this RNA-translational machine. This is just one example. Other sample systems include the proteins involved in the transcription process, RNAsomes that are involved in splicing activities, or protein systems involved in extracellular signalling. Although we have not reached or even come close to such sophistication in designing nanoscale devices, biological machineries provide design guidelines and inspirations for building nanoscale devices.

Mimicking biological systems for designing nanoscale devices may be a powerful strategy. The controlling factors in coordinating the 20 amino acids and 4 bases into functional units are non-covalent interactions. These interactions include hydrophobic interactions, van der Waals forces, hydrogen bonding, and molecular stacking. Nanoparticles, in general, are extremely dependent on molecular forces to assist them in maintaining their monodispersity. The ability to coordinate the molecular forces, similar to the ability of proteins to properly fold into a functional unit, on the surface of nanoparticles is difficult to do at this particular point since protein folding mechanisms remain somewhat unclear. A simpler and first step toward the use of biology to mediate nanoparticle assembly is the use of recognition biomolecules (RBs). RBs can be coated onto die surface of nanoparticles and be used to direct the formation of nanoparticle aggregates.

The simple mixing of antibody coated nanoparticles with their matching antigen in aqueous solution can lead to rapid aggregation of nanoparticles. The ordering of nanoparticles within the aggregate network can be coordinated by the RBs. For example, proteins such as avidin have four binding sites to the small organic molecule biotin as shown in Figure 1.3; antibodies are in homogeneous in structure and have three binding sites. These two protein systems will yield nanoparticle nanoparticle aggregates of different shape, size, and nanoparticle spacing. For nanoparticle assembly, proteins may one day act as "molecular glue" to join nanoparticles but will unlikely be used for building nanodevices that require dynamic assembly since protein-induced aggregation may be irreversible. Beyond network formation, the conjugation of proteins to the surface of nanoparticles provides with greater functional capabilities. Researchers in nanotechnology proposed the powering of inorganic nanodevices with biomolecular motor. They demonstrated that biomolecular motors such as F1-adenosine triphosphate synthase and myosin provide enough force to propel inorganic nanoparticles in solution. And another research team proposed the use of motor proteins with microtubule track systems to construct molecular conveyer belts to build nanoscale devices. They are mimicking the biological process of vesicle transport inside cells.

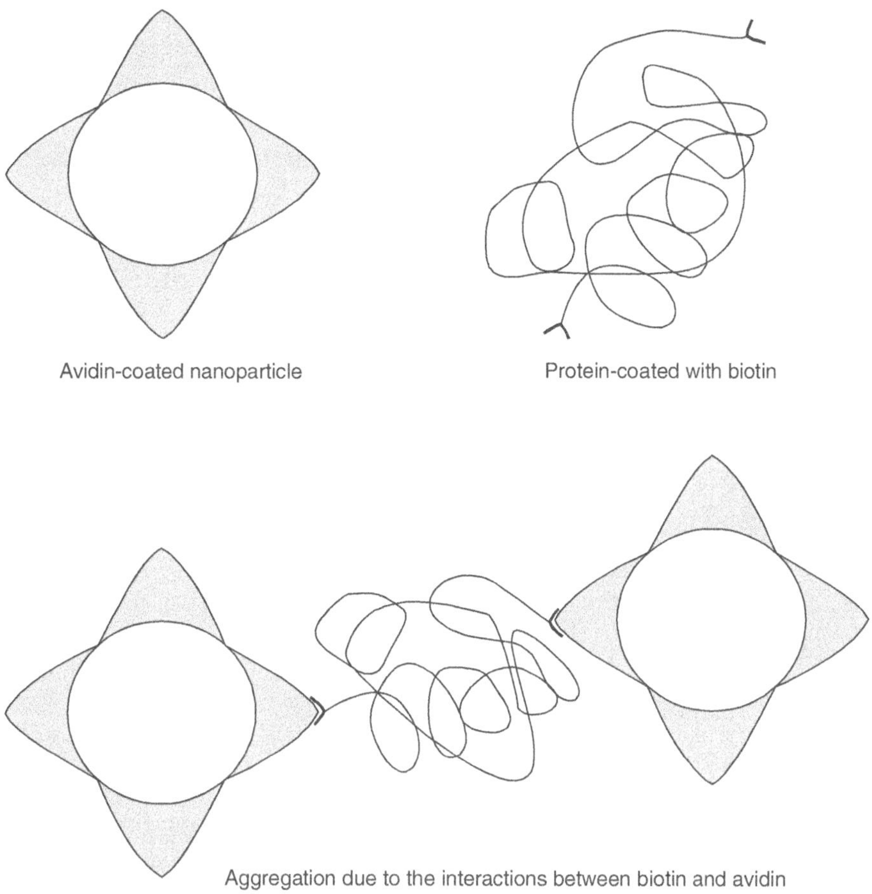

FIGURE 1.3 PROTEINS ARE DIRECTING THE ASSEMBLY OF NANOPARTICLES

CURRENT STATE AND HIGHLIGHT OF BIOAPPLICATIONS OF NANOSTRUCTURES

Although the successful design of nanoscale devices is several decades away, nanoparticles have been integrated with biological molecules for numerous biomedical applications. These can be considered, in some manners, to be simple, monofunctional nanostructures that can manipulate the release of drug molecules or to detect biomolecules in solution.

One of the most interesting concepts in functional nanoscale systems is the integration of thermosensitive hydrogels with colloidal nanoparticles for optically controlling drug release. Reports have already described the use of gold nanoshells for localized drug delivery. These nanoshells are made from a silica dielectric core with a gold shell, and their plasmon resonance wavelength can be tuned by varying the size of the core and the shell. It has been well known that colloidal gold nanoparticles and more recently, colloidal nanoshells have large absorption cross sections (e.g. 5-nm-diameter gold particle has a cross section 3 nm^2 at 514 nm). Because of this, colloidal gold nanoparticles and nanoshells possess a photothermal effect that is induced by nonradiative collisions between optically excited electrons with solvent molecules. Thermal-sensitive polymers such as the copolymer N-isopropylacrylamide and acrylamide can undergo a phase-change when the polymer is heated above their lower critical solution temperature (LCST). At a temperature > LCST, molecules trapped inside the polymer are released into the external environment due to the shrinking of the polymer. Essentially, molecules such as drug agents that are trapped in the core are squeezed out of the hydrogel when heated. The development of thermal-sensitive polymer matrices on the surface of gold nanoshells can be utilized for storage of drug agents while the gold nanoshells can act as an optical switch to control the release of drug agents. It is believed that one day, nanoparticle/drug-storing system can be directly targeted to lesions *in vivo*. The advantage is that the drug molecules are protected from the immune defence systems *in vivo*; and can be selectively released at sites of injury, which is expected to reduce side effects.

Recent, research has demonstrated the successful targeting of nanoparticles to specific sites in living animals, and peptides that specifically target tumour vasculatures have been discovered. Two of these peptides were conjugated onto the surface of two different-emitting quantum dots. These peptide-conjugated quantum dots were introduced into a mouse bearing xenograft tumour through injection into the tail vein. After 20 minutes, two distinct fluorescence signals were distinctly apparent on the tumour tissues upon optical excitation. Co-staining with markers snowed the localization of red-emitting quantum dots in the blood vessels while green-emitting quantum dots localized in the lymphatic vessels. No significant quantum dot emission was apparent in other nearby tissues and organs. Targeting molecules have been identified for tumour vessels and normal vessels in the brain, kidney, lungs, skin, pancreas, and other tissues. The integration of targeting molecules with nanostructures should provide a means of delivering nanostructures to specific cells and tissues *in vivo* as well as nanoparticle-based contrast agents for ultrasensitive optical imaging.

Beyond *in vivo* applications of nanostructures, interfacing organic molecules with inorganic nanostructures have led to the development of a new generation of *in vitro* detection systems. The unique absorption characteristics of aggregated and non-aggregated gold nanoparticles for the detection of genetic mutations was described by various research teams. Already a metallic barcoding system that can analyse thousands of biomolecules simultaneously has been developed. Colloidal metallic nanoparticles have also been utilized for genomic and proteomic screening. Researchers have developed highly selective electronic biosensors by using protein-adsorbed carbon nanotubes. Furthermore, there has been significant progress in the development of semiconductor quantum dots for biosensing and detection applications. Other types of nanoparticles such as fullerenes and dendrimers have found applications in drug storage and delivery. Magnetic nanoparticles are utilized as contrast agents for enhancing MRI imaging. Monofunctional nanostructures are rapidly advancing toward everyday use in research labs. In the future, we can foresee the development of multifunctional nanostructures where onset and evolution of a disease can be sensed by the nanostructure; the sensing of the disease, then, cause the selective release of one type or a combination of drug agents.

CONCEPT CHECK

1. Who proposed the term Nanotechnology?
2. What is the contribution of Richard P. Feynman?
3. Does a microbe have the potential to produce nanotubes? Mention the name.

MIND TWISTER

1. What is the difference between nanoparticles and dendrimers?

REFERENCES

Arnall, A.H. (2003). *Future Technologies, Today's Choices–Nanotechnology, Artificial Intelligence and Robotics; A technical, political and institutional map of emerging technologies.* Greenpeace Environmental Trust, London.

Chung, D. (2003). Nanoparticles have health benefits too. *New Scientist.* 16.

Dunn, S., and Whatmore, R.W. (2002). Nanotechnology advances in Europe. Working Paper STOA 108 EN, European Commission, Brussels.

Erhardt, D. (2003). Influence of ceramic surface conditioning and resin cements on microtensile bond strength to a glass ceramic. *Nat. Mater.* 2, 509.

Ferin, J. (1990). Increased pulmonary toxicity of ultrafine particles: I. Particle clearance, translocation, morphology *J. Aerosol Sci.* 21, 381.

Holister, P. and Harper, T.E. (2002). *The Nanotechnology Opportunity Report.* CMP Cientifica, Madrid.

Holister, P. (2002). *Nanotech–The Tiny Revolution.* CMP Cientifica, Madrid.

Iijima, (1991). S. High-yield fullerene encapsulation in single-wall carbon nanotubes. *Nature.*354, 56.

Jones, A. and Mitchell, M. (2001). *Nanotechnology: Commercial Opportunity.* Evolution Capital Ltd, London.

Kroto, H. W., et al. (1985). Electrochemical production of low-melting metal nanowires. *Nature.* 347, 354.

Masciangioli, T. and Zhang, W.X. (2003). A study of the origin of individual $PM_{2.5}$ particles in Shanghai air with synchrotron X-ray fluorescence microprobe. *Environ. Sci. Technol.* 37(5), A102.

Mossman, B. T. (1990). Free radical mechanisms in chemical pathogenesis : Summary of the symposium presented at the 1988 Annual Meeting of the Society of Toxicology. *Science.* 247, 294.

Murday, J. S. (2002). The Coming Revolution: Science and Technology of Nanoscale Structures. *The AMPTIAC Newsletter.* 6 (1), 5.

Nanoscience and nanotechnologies: opportunities and uncertainties. The Royal Society and The Royal Academy of Engineering, London, UK. (2004). (also available at www.nanotec.org.uk)

Nanotechnology: Views of the General Public, Report, BMRB International Ltd., London, UK. (2004). www.nanotec.org.uk

Nanophase materials innovative developments mold a thriving industry, D212. Frost and Sullivan, New York. (2001).

Oberdörster, (1996). G. A clearance model of refractory ceramic fibers (RCF) in the rat lung including fiber dissolution and breakage. *Inhal. Toxicol.* 8, 73.

Opportunities for Industry in the Applications of Nanotechnology, DTI Foresight Materials Panel Report, London. (2000).

Rittner, M. (2001). *Opportunities in Nanostructured Materials, GB-201.* Business Communications Co. Inc., Norwalk.

Román, C. *et al.* (2004). *Nanotubes.* Cientifica www.cientifica.com/html/docs/ Nanotubes. 2004_ExSum.pdf

Technology Alert-Nanotechnology, Institute of Nanomaterials, Glasgow. (2001).

The nanotech report 2003, Investment Overview and Market Research for Nanotechnology. Vol II, Lux Capital, New York (2003).

Wood, S., *et al.* (2003). *The Social and Economic Challenges of Nanotechnology.* EPSRC, Swindon.

Zhang, W. (2003). Nanoparticle suspension preparation using the arc spray nanoparticle synthesis system combined with ultrasonic vibration and rotating electrode. *J. Nanoparticle Res.* 5, 323.

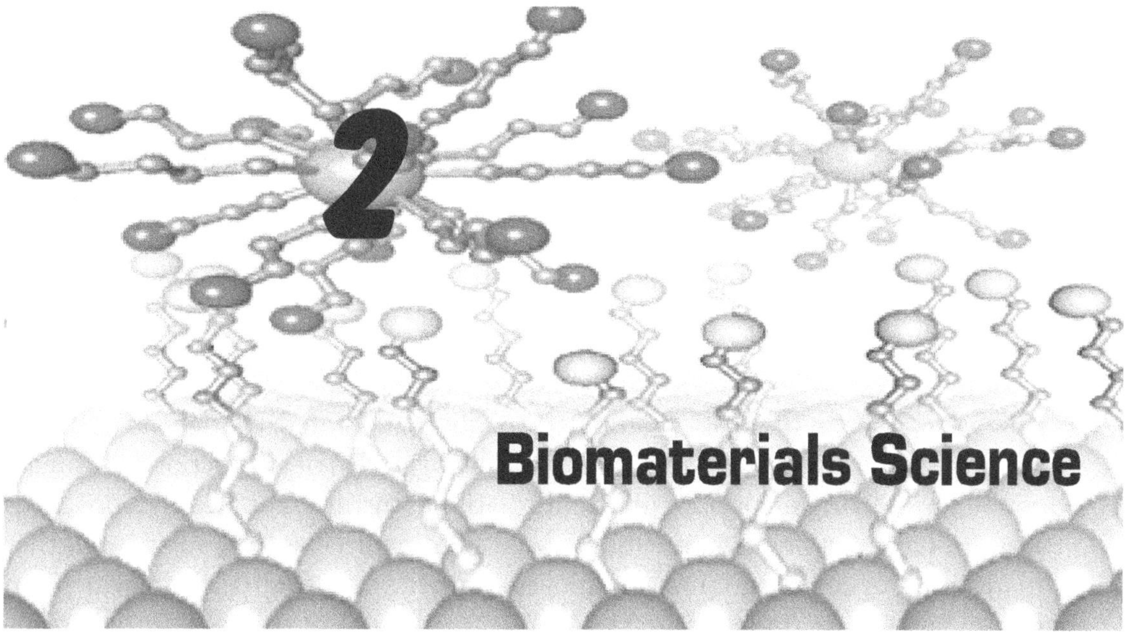

2 Biomaterials Science

BASIC INTRODUCTION ABOUT BIOMATERIALS

Human history has been famously subdivided into distinct ages of materials. Indeed, the progress of scientific sophistication has been tightly linked to the nature of the materials that were predominant in the different eras. The Stone Age, for example, represented the earliest of human ingenuity, when our primitive ancestors created and improved simple tools out of natural rocks. Subsequent Bronze and Iron Ages revealed increased technological savvy as tools and equipment were made from extracted and processed materials. In the Plastic Age, humankind perfected the science of organic synthesis and recently the Silicon Age has yielded unprecedented progress in extremely complex electrical devices. In the future, each of these materials will most likely occupy its own niche such that no new material will completely render the previous ones obsolete. However, it can be argued that further progress in particular fields such as regenerative medicine, engineering and recently nanotechnology will benefit tremendously from the discovery and development of novel, designed and integrated materials.

Biomaterials is an exciting field with steady growth, which encompasses aspects of medicine, biology, chemistry and materials science. "A biomaterial is a nonviable material used in a medical device, intended to interact with biological systems". Biomaterials science is the physical and biological study of materials and their interaction with the biological environment. Traditionally, the most intense development and investigation has been directed towards biomaterials synthesis, optimization, characterization, testing and the biology of host–material interaction."

Over the last two centuries, the field of medicine has increasingly utilized biomaterials in the investigation and treatment of disease. Common examples include surgical sutures and needles, catheters, orthopaedic hip replacements, vascular grafts, implantable pumps, and cardiac pacemakers. The purpose of this chapter is to provide a historical overview of biomaterials, emphasizing the evolution of three generations of materials over the last century, and detailing current trends in the development and application of biomaterials in medicine.

Most have defined biomaterials broadly. Park stated that a *biomaterial* is "a synthetic material used to replace part of a living system or to function in intimate contact with living tissue." Similarly, the National Institutes of Health Consensus Development Conference of 1982 defined a biomaterial as "any substance other than a drug, or combination of substances, synthetic or natural in origin which can be used for any period of time as a whole or as a part of the system that treats, augments, or replaces any tissue, organ or function of the body." In contrast, a *biological material* is a substance produced by a living organism. Muscle and bone are examples. But, synthetic materials that only come in contact with the skin, such as hearing aids and bandages, are generally not categorized as biomaterials.

Considerable current effort is directed towards the development of engineered surfaces that could elicit rapid and highly precise reactions with cells and proteins, tailored to a specific application. Indeed, a complementary definition essential for understanding the goal (i.e., specific end applications) of biomaterials science is that of "biocompatibility." Biocompatibility is the ability of a material to perform with an appropriate host response in a specific application. Examples of "appropriate host responses" include the resistance to blood clotting, resistance to bacterial colonization, and normal, uncomplicated healing. Examples of specific applications include a haemodialysis membrane, a urinary catheter or hip-joint replacement prosthesis. This general concept of biocompatibility has been extended recently in the broad approach called "tissue engineering: in which *in vitro* pathophysiological processes are harnessed by careful selection of cells, materials and metabolic and biomechanical conditions to generate functional tissue.

Advances in medicine, surgery, biotechnology, and material science have significantly influenced the development and application of biomaterials. Specifically, advances in surgical technique and instrumentation have enabled the placement of implants into previously poorly accessible locations. Examples include the placement of endovascular stents and the surgical insertion of mechanical heart valves. Similarly, advances in biotechnology have led to the development of scaffolds for tissue engineering. Today, biomaterials can be found in over 8000 different types of medical devices. Table 2.1 lists several current applications of biomaterials in modern medicine.

TABLE 2.1 USES OF BIOMATERIALS

Problem area	Examples
Replace diseased or damaged parts	Soft or hard tissue prosthetic implants, cardiac valve replacements, renal dialysis machines, tissue engineering scaffolds
Assist in healing	Sutures, adhesives and sealants, bone screws and nails, plates
Improve function	Cardiac pacemakers, intraocular lens
Correct functional abnormality	Cardiac defibrillator/pacemaker
Correct cosmetic problem	Soft tissue implants (breast, chin, calf)
Aid diagnosis of disease	Probes, catheters and biosensors
Aid treatment of disease	Catheters, drains, implantable pumps, and controlled drug delivery systems

HISTORICAL BACKGROUND

The first reported clinical application of a "biomaterial" can be traced back to 1759, when Hallowell repaired an injured artery using a wooden peg and twisted thread. However, it was not until the promotion of aseptic surgical techniques in the 1860 by Lister that the practical use of biomaterials became possible. Before this, surgical procedures were often complicated by serious and often life-threatening infection. Foreign materials deliberately implanted into the body exacerbated this problem, often representing for an infection that the body's natural immune response could not effectively penetrate. As the widespread implementation of sterile techniques brought infection rates under control, the impact of the physical properties of specific medical materials on the success of implant procedures was recognized.

The recognition of the therapeutic potential of biomaterials, along with advances in surgical techniques, led to increasing interest in the incorporation of synthetic materials into living tissues. These early implants were largely applied to the skeletal system. In the 1900s, bone plates were used to fix long bone fractures, though many of these early plates failed as a result of poor mechanical design. Also, early materials chosen primarily for their mechanical properties often corroded rapidly in the body and inhibited the natural healing processes. The early use of vanadium steel in orthopaedic implants is an example of this.

Despite these early troubles, improved designs and more suitable materials were soon introduced. In the 1930s, the introduction of stainless steel and cobalt–chromium alloys led to greater success in fracture fixation and the performance of the first joint replacement surgeries. Similarly, during World War II, it was noted that retained fragments of plastic from aircraft canopies did not result in chronic adverse reactions in injured pilots. Consequently, the use of plastics and polymers as biomaterials grew exponentially.

FIRST-GENERATION BIOMATERIALS (1950s–1960s)

GENERAL CHARACTERISTICS

The experiences of the late nineteenth and early twentieth centuries led to the development of the first set of modern biomaterials in the 1950s and 1960s. These "first-generation" biomaterials were specifically designed for use inside the human body and saw application in multiple disciplines of medicine including orthopaedics, cardiovascular surgery, ophthalmology, and wound healing. First-generation biomaterials were often based on commonly available materials selected from engineering practice. For example, the original dialysis tubing was made of cellulose acetate, a commodity plastic, and early vascular grafts were made of Dacron, a polymer developed in the textile industry. Though these materials facilitated the treatment of disease, it became clear that they had the potential to elicit serious inflammatory reactions. Therefore, newer materials were selected for two fundamental characteristics: the ability to be tolerated by the body and the ability to reproduce the natural functions of the tissues to be augmented or replaced.

The primary characteristic is often termed *biocompatibility* and refers to the acceptance of an artificial implant by the surrounding tissues and by the body as a whole. A completely biocompatible material would not cause thrombogenic, toxic, or inflammatory responses when placed into living tissue. Furthermore, the material would not elicit carcinogenic, mutagenic, or teratogenic effects. While biocompatibility remains a desired characteristic for all biomaterials, it should be noted that no synthetic material is completely biologically inert. Biocompatibility is more accurately a relative term.

The second characteristic, sometimes termed *biofunctionality*, refers to a biomaterial's ability to exhibit adequate physical and mechanical properties to augment or replace body tissues. As expected, these properties vary greatly depending on the target tissue. For example, a material being used for bone augmentation must exhibit a high compressive strength, while a material used for ligament replacement must exhibit a high degree of flexibility and tensile strength. Finally, for practical purposes, a biomaterial must be amenable to being machined or formed into different shapes, have relatively low cost, and be readily available.

In general, biomaterials are categorized by their origin (i.e., natural or synthetic) as well as by their chemical composition (i.e., polymers, ceramics, alloys, and composites). The following sections provide an overview of various first-generation biomaterials and their predecessors. However, it should be noted that these categories may also apply to newer generations of biomaterials and that there is significant overlap between the three generations of biomaterials highlighted in this chapter.

Naturally Occurring Biomaterials

Naturally occurring biomaterials encompass biological products of nonhuman origin that are or were used in clinical applications. For example, cellulose, catgut, ivory, silk, natural rubber, glass, graphite, and several pure metals are natural biomaterials. Historically, natural biomaterials were some of the first devices to be used in clinical practice. In 1860, Lister reported the use of catgut as a suture material. Similarly, natural rubber was used by Horsley in the early 1900s for the development of synthetic grafts. Today, naturally occurring biomaterials have largely been replaced by synthetic materials deliberately designed with specific characteristics.

Metals and Alloys

Metals are commonly used for load-bearing implants. Specifically, orthopaedic procedures utilize a variety of metals to replace or augment skeletal function. Examples range from simple plates and screws to complex joint prostheses. In addition, metals are used in cardiovascular surgery, and for dental and maxillofacial implants. Overall, the biocompatibility of metallic implants is an important characteristic because these implants can corrode in the *in vivo* environment. Corrosion leads to the disintegration of the implant material and the release of potentially harmful products into the surrounding tissues.

Pure metals The noble metals, such as gold, silver, and platinum, represent some of the earliest used biomaterials due to their tissue compatibility and corrosion resistance. For example, metal ligatures of silver, gold, platinum, and lead were utilized by Levert in 1829. Similarly, Cushing used silver clips in 1911 to control bleeding during operations to remove cerebral tumours. While the pure metals were the earliest metals used as biomaterials, they have been steadily replaced by alloys engineered for improved strength and biocompatibility. In fact, titanium, which was initially used in World War II for aircraft devices, has been the only new pure metal biomaterial introduced since 1940.

Alloys Metal alloys have largely replaced pure metals as biomaterials. Although many alloys are used in medical devices, the most commonly employed are stainless steel, commercial titanium alloys, and cobalt-based alloys. The first metal alloy developed specifically for human use was "Vanadium steel," which was used to manufacture bone fixation plates and screws in the early 1900s. However, as mentioned earlier, Vanadium steel corrodes *in vivo*. Since then, several stainless steel alloys have been developed with greater strength and improved corrosion resistance. These include 18-8 or type 302 stainless steel as well as the later 18-8sMo or type 316 stainless steel. 18-8sMo steel was unique in that it contained a small percentage of molybdenum to improve its corrosion resistance in salt water. Reduction of the carbon content from 0.08 to 0.03% led to the development of type 316L steel. Together, types 316 and 316L stainless steel are known as the austenitic stainless steels and are the most widely used alloys for biomaterial applications.

Similar to stainless steel alloys, cobalt-chromium alloys were developed for commercial application in the early 1900s and subsequently utilized as biomaterials. These alloys were used as alternatives to gold alloys in dentistry and have recently been utilized in the production of artificial joints. Most recently, titanium and its alloys have been used as biomaterials for implant fabrication. Initially discovered by Gregor in 1791, titanium remained a laboratory curiosity until 1946, when Kroll developed a process for the commercial production of titanium by reducing titanium tetrachloride with magnesium. Since then, titanium and its alloys have been widely used as biomaterials because of their relative biological inertness and superior mechanical properties. Specifically, titanium is a reactive metal that forms a tenacious oxide layer on its surface when exposed to air, water, or specific electrolytes. This oxide layer provides a protective coating that shields the material from chemical degradation and the biological environment. Conversely, the oxide layer, which is in contact with body tissues, is essentially insoluble and does not release ions that can react with other molecules. Lastly, titanium exhibits a high-yield strength and low elastic modulus, properties that are desirable in orthopaedic implants.

Shape-memory alloys (SMAs) Shape-memory alloys are a group of metals that have the interesting properties of thermal shape memory, superelasticity, and force hysteresis. Nitinol, an approximately equiatomic alloy of nickel and titanium, is the most widely used member of this group. Originally discovered at the U.S. Naval Ordinance Laboratory and then reported by Beuhler and colleagues in 1963, nitinol is relatively biocompatible and more compliant than most other alloys. Currently, nitinol is used in an increasing number of surgical prostheses and disposables, including a variety of endovascular stents, intracranial aneurysm clips, and vascular suture anchors.

CERAMICS

Ceramics are one of the oldest artificially produced materials, used in the form of pottery for thousands of years. Ceramics are polycrystalline compounds including silicates, metallic oxides, carbides, and various refractory hydrides, sulphides, and selenides. The first clinical application of ceramics was the use of plaster of Paris as a casting material. However, until recently, the general use of ceramics as implantable biomaterials was limited due to their inherent brittleness, low tensile strength, and low impact strength. In recent years, newer, "high-tech" ceramics have gained increased use as biomaterials due to their relative bioinertness and high compressive strength. These implantable ceramics have been termed *bioceramics* and are grouped into three categories based on their biological behaviour in certain environments: the relatively bioinert ceramics, the bioreactive or surface-reactive ceramics, and the biodegradable or reabsorbable ceramics. Relatively bioinert bioceramics are nonabsorbable carbon-containing ceramics, alumina, zirconia, and silicon nitrides. While in a biological host, relatively bioinert ceramics maintain their mechanical and physical properties. They are used in dense and porous forms, usually have good wear, and are excellent for gliding functions.

For example, bioinert bioceramics are used to produce femoral head replacements. In addition, relatively bioinert materials are typically used as structural-support implants such as bone plates and bone screws. The bioactive ceramics include glass, glass-ceramics, and calcium phosphate-based materials. They are characterized by their ability to provoke surrounding bone and tissue responses, which makes them advantageous for anchoring an implant or reducing its stress. They have been successfully used as coatings, continuous layers, and embedded particles in orthopaedics and dental surgery.

Lastly, biodegradable or resorbable ceramics include aluminium calcium phosphate, coralline, plaster of Paris, hydroxyapatite, and tricalcium phosphate. They differ from the bioactive ceramics in two major ways. First, they are more soluble, and consequently are degraded by surrounding tissues. Second, due to their porous structure they may stimulate tissue ingrowth and therefore offer the potential to fill or bridge defects. These materials have been used in the fabrication of various orthopaedic implants as well as for solid or porous coatings on hybrid implants made of other biomaterials.

Overall, bioceramics are unique in their ability to form porous structures. This is advantageous because a large surface area-to-volume ratio results in a greater tissue contact surface. In addition, interconnected pores permit tissue ingrowth and may facilitate blood and nutrition delivery. Though the bioactive and biodegradable ceramics are often classified as second-generation biomaterials due to their dynamic qualities.

Polymers

In the 1890s, the earliest synthetic plastics were developed using cellulose, a major structural component of plants. In the 1930s, nylon, the first commercial polymer, was produced and made widely available, leading to the birth of the field of polymer science. However, the field of polymer science was relatively limited until 1954, when Professor G. Natta developed a new polymerization technique that transformed random structural arrangements on noncrystallizable polymers into structures of high chemical and geometrical regularity. This discovery had a significant impact on the development of propylene polymers, an inexpensive petroleum derivative used for a variety of medical applications.

Since then, synthetic polymeric materials have been extensively used in a variety of biomedical applications including medical disposables, prosthetic materials, dental materials, implants, dressings, and extracorporeal devices. Overall, polymeric biomaterials display several key advantages. These include ease of manufacturing into products with a wide variety of shapes, ease of secondary processability, reasonable cost, wide availability, and wide variety of mechanical and physical properties. Polymers consist of small repeating units, or isomers, strung together to form long-chain molecules. These long chains are formed by covalent bonding along the backbone chain and can be arranged into linear, branched, and network structures, depending on the functionality of the repeating units. The long-chains may be held together by a variety of chemical and ionic forces. These include secondary bonding forces, such as van der Waals forces and hydrogen bonds, and primary covalent bonding forces through cross links between chains. Such cross-linking can increase the density of materials to improve their strength and hardness. However, cross-linked materials often lose their flexibility and become more brittle. The physical properties of polymers can be deliberately changed in many ways.

In particular, altering the molecular weight and its distribution has a significant effect on the physical and mechanical properties of a polymer. For example, with increasing molecular weight, the chains of a polymer become longer and less mobile, resulting in a more rigid material. Similarly, changing the chemical composition of the backbone or side chains can change the physical properties of the polymer. For example, the substitution of a backbone carbon in a polyethylene with divalent oxygen increases the rotational freedom of the chain, resulting in a more flexible material. Likewise, side chain substitution, cross-linking, and branching all affect the physical properties of polymers. Increasing the size of side groups or branches, or increasing the cross-linking of the main chains all result in a poorer degree of molecular packing. This retards the polymer crystallization rate, thereby decreasing the melting temperature of a material. Similarly, changes in temperature can have a significant effect on the properties of polymers. The *glass transition temperature* refers to the point between the temperature range in which a polymer is relatively stiff (glassy region), and the temperature range in which a polymer is very compliant (rubbery region). Depending on the temperature, a single polymeric material can take on a variety of forms.

Today, a variety of polymers are used as biomaterials. These include polyvinylchloride (PVC), polyethylene (PE), polypropylene (PP), polymethylmethacrylate (PMMA), polystyrenes (PS), fluorocarbon polymers (most notably polytetrafluoroethylene or PTFE), polyesters, polyamides or nylons, polyurethanes, resins, and polysiloxanes (silicones). These synthetic polymers have found extensive utilization in a wide variety of applications including implantable devices, coatings on devices, catheters and tubing, vascular grafts, and injectable drug delivery and imaging systems.

Composites

Composite biomaterials are composed of two or more distinct constituent materials, incorporating the desired physical and mechanical properties of each. This results in a hybrid product whose overall properties may be significantly different from the homogeneous materials. For example, rubber used in various catheters is often filled with very fine particles of silica to enhance the strength and toughness of the material. Other examples of composite biomaterials in clinical use include composite resins commonly used as dental fillings (e.g. polymer matrix and barium, glass, or silica inclusions) as well as porous orthopaedic and soft-tissue implants.

SECOND-GENERATION BIOMATERIALS (1970s–2000)

General Characteristics

Over the last three decades, the field of biomaterials shifted away from the traditional properties of first-generation biomaterials. While biocompatibility and biofunctionality continued to remain important, the long-standing goal of achieving a bioinert tissue response began to be replaced by the notion of developing materials that were *bioactive,* or *biodegradable*. These bioactive or biodegradable biomaterials have been termed "second-generation" biomaterials and have seen increasing clinical application. While bioinert materials were designed to elicit little or no tissue response, bioactive materials have been designed to elicit specific and controlled interactions between the material and the surrounding tissue. For example, synthetic hydroxyapatite (HA) ceramics are used as porous implants, powders, and coatings on metallic prostheses to provide bioactive fixation. HA coatings on implants lead to a controlled tissue response in which bone grows along the coating. This response, termed osteoconduction, promotes the formation of a mechanically strong interface between the implant and the native bone. Such materials have seen widespread clinical use in the fields of orthopaedics and dentistry. Overall, bioactivity is increasingly becoming a characteristic of modern biomaterials as expanding applications call for dynamic biomaterials and devices.

Similarly, biodegradable or bioresorbable materials were designed to exhibit clinically relevant breakdown and absorption. In this manner, the issue of interface integration of the implant and surrounding tissue is addressed as the foreign material is eventually replaced by regenerating tissues. A common example is absorbable sutures. Consisting of a polymer composed of polylactic acid (PLA) and polyglycolic acid (PGA), these sutures decompose into carbon dioxide and water after a designated length of time. Consequently, the foreign material used to approximate tissue during surgery is resorbed after the tissue has naturally healed together. Other applications of bioresorbable materials include resorbable fracture fixation plates and screws as well as controlled-release drug delivery systems. The following sections detail several classes of second-generation biomaterials, including biodegradable polymers, hydrogels, and bioactive and biodegradable ceramics.

Biodegradable Polymers

Many types of biomaterials utilized for soft- and hard-tissue repair are required only for a short time to support tissue regeneration and ingrowth. These temporary biomaterials have been constructed using biodegradable polymers that degrade when placed in the body, allowing functional tissue to grow in its place. The mechanisms for degradation for these types of polymers include both hydrolytic and enzymatic degradation. These degradation processes are influenced by many factors including molecular weight and distribution of the polymer, chemical composition of the polymer backbone, polymer morphology (e.g. amorphous/crystalline structure), glass transition temperature, additives, and environmental conditions (pH, temperature, etc.).

Biodegradable polymers can be natural or synthetic in origin. Natural polymers include starch, chitin, collagen, and glycosaminoglycans. For example, preparations of chitin and its derivative chitosan have been developed for wound dressings and sustained-release drug delivery systems. The first totally synthetic biodegradable polymer was polyglycolic acid which was introduced in the early 1970s. This was followed by the polylactides in 1985. Both of these polymers fall under the poly (α-hydroxy acid) family of polymers, which is the most widely used for the production of biodegradable biomaterials. Other biodegradable polymers include polycaprolactone, poly(orthoesters), polyanhydrides, and polyphosphazenes.

Currently, biodegradable polymers are used for a variety of biomaterials applications including sutures, wound dressings, fracture plates and screws, and controlled release delivery systems. In addition, biodegradable matrices and scaffolds show a great deal of promise in the field of tissue engineering due to their ability to degrade while encouraging functional host tissue to take their place.

Hydrogels

Hydrogels are cross-linked hydrophilic polymer networks that can absorb water or other biological fluids. First used for a biomedical application in the late 1950s as a soft contact lens material, hydrogels have since found widespread application as orthopaedic and ophthalmic implants, controlled drug systems, and biosensors. Even though hydrogels were first utilized in the 1950s, they are included here as second-generation biomaterials because of their unique properties. Hydrogels display several modifiable characteristics that make them useful as biomaterials. These characteristics, which are primarily determined by the hydrogel's polymer network structure, include a hydrogel's swelling behaviour, diffusive characteristics, and surface properties. The swelling behaviour of a hydrogel refers to the material's ability to absorb water or other fluids and is largely a function of its structural network properties. When a biopolymer network comes in contact with an aqueous solution, the network starts to swell due to the interaction of the polymer chains and water. This propensity to swell is offset by the resistive force induced by the cross links of the polymer network. When these two forces balance, equilibrium is reached and swelling stops. In addition to the network structure of the polymer, a hydrogel's swelling behaviour can be influenced by the presence of ionizable pendant groups. In such cases, the forces influencing the swelling equilibrium may be altered by the localization of charges within the hydrogel.

The diffusive characteristics of a hydrogel have been the basis of many of their newer applications, including drug and protein delivery. While the chemical potential gradient of a solute determines its flux within a system, several important characteristics of hydrogels may serve to influence the rate and pattern of this diffusion. These include the structure and pore size of a hydrogel, the polymer composition, the water content, and the nature and size of the solute. In addition, interactions between different solutes, between solutes and the gel polymers, and between solutes and solvents all play a role in determining the overall diffusion characteristics of a solute and a hydrogel. Finally, the surface properties of hydrogels may be altered to achieve a variety of unique characteristics. In general, hydrogels have a complex surface structure composed of numerous dangling chains attached on one end to the polymer network. By altering these chains, the surface properties of hydrogels can be engineered to serve a number of different purposes. Hern and Hubbell incorporated adhesion-promoting oligopeptides into hydrogels, giving them the ability to mediate cell adhesion properties. Similarly, by using modified lipid bilayers, hydrogel surfaces have been engineered to mimic cell membranes.

By adjusting these unique characteristics, engineered hydrogels have recently been created to serve a variety of biomedical applications. Specifically, controlled drug delivery systems have been developed by altering the diffusive characteristics of hydrogels. Such systems are able to maintain a desired blood concentration of a specific drug for a prolonged period of time and have been constructed as either matrix or reservoir configurations. In a reservoir system, the agent is stored in a central core surrounded by a polymer membrane. Conversely, in a matrix system, the agent is uniformly distributed throughout the material and slowly released from it. In addition, these drug delivery systems can be constructed using biodegradable polymers, thereby negating the necessity of surgical removal.

More recently, hydrogel-based delivery systems have been created that release their agents in response to changes in their environment. These new, bioactive systems may respond to changes in pH, temperature, ionic potentials, solvent concentrations, and magnetic or electrical fields. The use of such systems is being investigated for the controlled release of a variety of agents including insulin, streptokinase, lysozyme, and salmon calcitonin. In addition, a recently recognized advantage of hydrogels is that they may have the ability to protect embedded drugs, peptides, or proteins from the potentially harsh biological environment. This may enable the oral delivery of engineered proteins, which may otherwise be prematurely broken down in the digestive tract.

Bioactive and Biodegradable Ceramics

As stated earlier, the bioactive ceramics include glass, glass-ceramics, and calcium phosphate-based materials. They are characterized by their ability to provoke surrounding bone and tissue responses, which makes them advantageous for anchoring an implant or reducing its stress. The biodegradable ceramics include aluminium calcium phosphate, coralline, plaster of Paris, hydroxyapatite, and tricalcium phosphate. They are unique in that they are soluble and are therefore degraded and absorbed by surrounding tissues. In addition, due to their porous structure they stimulate bone ingrowth. These materials have been used in the fabrication of various orthopaedic implants, as well as for solid or porous coatings on implants made of other biomaterials.

THIRD-GENERATION BIOMATERIALS (2000 ONWARDS)

Recently described by Hench and Polak, a new, third-generation of biomaterials is being designed. For this class of biomaterials, the distinct characteristics of bioactivity and biodegradability have now converged. As opposed to second-generation biomaterials, which are generally either bioactive or biodegradable, third-generation biomaterials are being designed that display both of these characteristics. Furthermore, these new biomaterials are being designed alongside advances in the fields of tissue engineering, microfabrication, and nanofabrication. In this manner, third generation biomaterials will be created to aid in the regeneration, and not simply the replacement of injured or lost tissues. In addition, micro- and nanofabrication techniques are being utilized to create "smart" biomaterials and implants that can detect and respond to various tissue and cellular stimuli.

Types of Biomaterials

Biodegradable biomaterials Biodegradation is the process of breaking down into small particles by biological means. Due to biodegradation, insoluble solid polymers tend to reduce to soluble fragments that are either excretable or metabolized under some physiological conditions. During surgery, desirability of the polymeric device which introduced into the body does not need to be retrieved throughout the lifetime. For example, introduced polymeric biomaterials will be there in our body to fill and support the bone defect till the natural bone grows back or to provide the delivery of drug, until a condition is corrected.

Bioeliminable biomaterials Generally big molecules are difficult to eliminate or excrete via any of the metabolic pathways. But under some physiological conditions, such molecules can be dissolved into low molecular weight compounds and be eliminated/excreted very easily. For example, polymeric compound of polyalkylcyanoacrylate is very difficult to eliminate due to the presence of alkyl group in the compound, which renders a hydrophobic environment to the polymer. Here the physiological conditions which convert the alkyl group, make the compound hydrophilic in nature, thereby enabling them to get eliminated easily. Dextran and polyethyleneglycol or polyethylene oxides are good examples for bioeliminable biomaterials.

Bioabsorbable biomaterials The biomaterials introduced to the body for the substitution purpose, is sometimes permanently utilized by the body itself. Thus biomaterials are known as bioabsorbable biomaterials. Suitable example is calcium phosphate. It is frequently used by dentists to fill the gap in between the teeth, which is going to be utilized for a longer period.

Removable or non-biodegradable biomaterials Teflon is a good example for non-biodegradable biomaterials because of its high susceptibility nature. This can be used as cardiovascular graft and in preparation of stents.

Overview about Biopolymers

Biopolymers or biodegradable polymers is a newly emerging field. A vast number of biodegradable polymers have been synthesized recently and some microorganisms and enzymes capable of degrading them have been identified. In developing countries, environmental pollution by synthetic polymers has assumed dangerous proportions. As a result, attempts have been made to solve these problems by including biodegradability into polymers that have everyday use, through slight modifications of their structure.

Biodegradation is a natural process by which the organic chemical in the environment is converted to simpler compounds, mineralized and redistributed through elemental cycles such as carbon, nitrogen and sulphur cycles. Biodegradation can only occur within the biosphere as microorganisms play a central role in the biodegradation process. Biopolymers are polymers formed in nature during the growth cycles of all organisms; hence, they are also referred to as natural polymers. Their synthesis generally involves enzyme-catalysed, chain growth polymerization reactions of activated monomers, which are typically formed within cells by complex metabolic processes.

Some examples are as follows:

1. Polysaccharides—starch, cellulose, chitin, chitosan, alginic acid.
2. Polypeptides of natural origin—gelatin.
3. Bacterial polyesters
4. Polymers with hydrolysable backbones—polyesters, polycaprolactone, polyamides, polyurethanes and polyureas, polyanhydrides, poly(amide-enamine)s.
5. Polymers with carbon backbones—poly(vinyl alcohol) and poly(vinyl acetate), polyacrylates.

Cellulose

Many polymer researchers are of the opinion that polymer chemistry had its origins with the characterization of cellulose. Cellulose was isolated for the first time some

150 years ago. Cellulose differs in some respects from other polysaccharides produced by plants, the molecular chain being very long consisting of one repeating unit (cellobiose). Naturally, it occurs in a crystalline state. From the cell walls, cellulose is isolated in microfibrils by chemical extraction. In all forms, cellulose is a very highly crystalline, high molecular weight polymer, which is infusible and insoluble in all but the most aggressive, hydrogen bond-breaking solvents such as N-methylmorpholine-*N*-oxide. Because of its infusibility and insolubility, cellulose is usually converted into derivatives to make it more processable.

The biodegradation of cellulose is complicated, because cellulose exists together with lignin, for example, in wood cell walls. White-rot fungi secrete exocellular peroxidases to degrade lignin preferentially and, to a lesser extent, cellulases to degrade the polysaccharides in order to produce simple sugars which serve as nutrients for these microorganisms. Brown-rot fungi secrete enzymes for the degradation of cellulose and the hemicelluloses. Soft-rot fungi also degrade principally these two types of polysaccharides.

Recently developed series of cellulose acetate films, differing in their degree of substitution and that was evaluated in this bench-scale system. In addition, commercially available biodegradable polymers such as poly(hydroxybutyrate-co-valerate) (PHBV) and polycaprolactone (PCL) which were included as points of reference. Based on film integration and film weight loss, cellulose acetates, having degrees of substitution less than approximately 2.20, compost at rates comparable to that of PHBV. NMR and GPC analyses of composed films indicate that low molecular weight fractions are removed preferentially from the more highly substituted and slower degrading cellulose acetates.

Polyesters

Almost the only high molecular weight compounds shown to be biodegradable are the aliphatic polyesters. The reason for this is the extremely hydrolysable backbone found in these polyesters. Polyesters derived from diacids of medium-sized monomers are readily degraded by the fungi, *Aspergillus niger* and *Aspergillus flavus*. In order for synthetic polymers to be biodegradable by enzyme catalysts, the polymer must be able to fit into the enzyme's active site. This is the main reason why flexible aliphatic polyesters are degradable and the rigid aromatic polyesters are not. Poly(glycolic acid) (PGA) is the simplest linear, aliphatic polyester.

PGA and Poly(glycolic acid-co-lactic acid) (PLGA) are used as degradable and absorbable sutures. Their great advantage is the degradability by simpler hydrolysis of their ester backbone. Furthermore the degradation products are utilized, metabolized to carbon dioxide and water or are excreted via the kidney.

CHEMISTRY OF BIODEGRADABLE SOLID POLYMERS

Since a degradable implant does not have to be removed surgically once it is no longer needed, degradable polymers are of value in short-term applications that require only the temporary presence of a device. An additional advantage is that the use of degradable implants can circumvent some of the problems related to the long-term safety of permanently implanted devices. A potential concern relating to the use of degradable implants is the toxicity of the implant's degradation products that are released into the body of the patient. Thus the design of a degradable implant requires careful attention and needs testing for potential toxicity of the degradation products.

The term chemical degradation refers to a chemical process resulting in the cleavage of covalent bonds. Hydrolysis is the most common chemical process by which polymers degrade, but degradation can also occur via oxidative and enzymatic mechanisms. It is now widely believed that the chemical degradation of the polymeric backbone of poly(lactic acid) is predominantly controlled by simple hydrolysis and occurs independently of any biological agent.

Several distinct types of chemical degradation mechanisms have been mentioned clearly in Figure 2.1. Chemical reaction can lead to cleavage of cross links between water-soluble polymer chains (mechanism I), to the cleavage of polymer side chains resulting in the formation of polar or charged groups (mechanism II), or to the cleavage of the polymer backbone (mechanism III). Obviously combinations of these mechanisms are possible: for instance, a cross-linked polymer may first be partially solubilized by the cleavage of cross links (mechanism I), followed by the cleavage of the backbone itself (mechanism III). It should be noted that water is the key to all of these degradation schemes. Even enzymatic degradation occurs in the aqueous environment.

Since the chemical cleavage reactions described above can be mediated by water or by biological agents such as enzymes and microorganisms, it is possible to

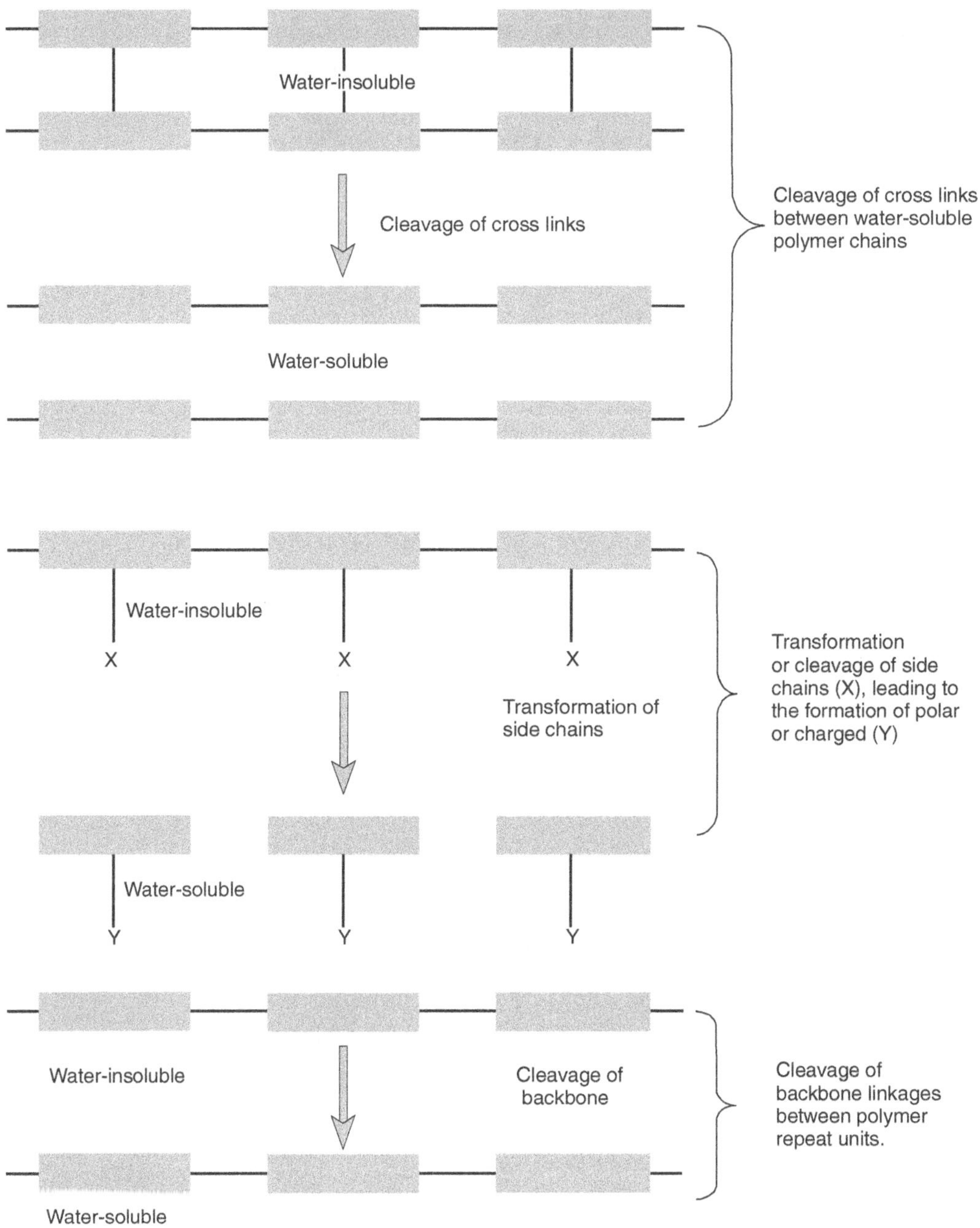

FIGURE 2.1 CHEMISTRY OF BIODEGRADABLE SOLID POLYMERS

distinguish between hydrolytic degradation and biodegradation, respectively. It has often been stated that the availability of water is virtually constant in all soft tissue and varies little from patient to patient. On the other hand, the levels of enzymatic activity may vary widely not only from patient to patient but also between different tissue sites in the same patient. Thus polymers that undergo hydrolytic cleavage tend to have more predictable *in vivo* rates than polymers whose degradation is mediated predominantly by enzymes. The latter polymers tend to be generally less useful as degradable medical implants.

MODES OF EROSION

When reviewing the literature, no clear distinctions in the meaning of the prefix "bio" is established, leading to the often interchangeable use of the terms "erosion" and "bioerosion". Although efforts have been made to establish the generally applicable and widely accepted definitions for all aspects of biomaterials research, there is still significant confusion even among experienced researchers in the field as to the correct terminology of various degradation processes.

The term "erosion" refers often to physical changes in size, shape, or mass of a device, which could be the consequence of either degradation or simply dissolution. Thus, it is important to realize that erosion can occur in the absence of degradation, and degradation can occur in the absence of erosion. A sugar cube placed in water erodes, but the sugar does not chemically degrade. Likewise, the embitterment of plastic when exposed to UV light is due to the degradation of the chemical structure of the polymer and takes place before any physical erosion occurs. In agreement with Heller's suggestion, we define a "bioerodible polymer" as a water-insoluble polymer that is converted under physiological conditions into water-soluble material(s) without regard to the specific mechanism involved in the erosion process. "Bioerosion" includes therefore both physical processes (such as dissolution) and chemical processes (such as backbone cleavage). Here "Erosion" indicates that the degradation is caused by high temperature, strong acids or bases, UV light or weather conditions. There are two modes of erosion which play an important role, and termed as erosion surface and bulk erosion. In surface erosion, degradation happens from exterior only with little or no water penetration into bulk. In bulk erosion, water penetrates the entire structure and degrades entire device simultaneously. These two erosions are represented in Figure 2.2.

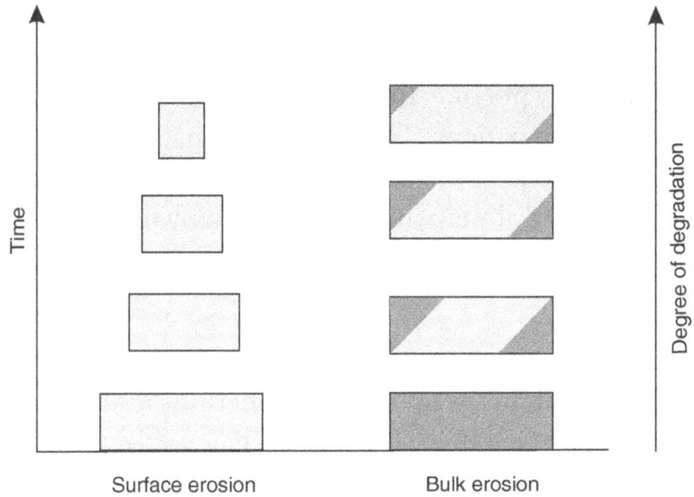

FIGURE 2.2 MODES OF EROSION

MOLECULAR STRUCTURE EFFECTS ON HYDROLYTIC BREAKDOWN

Hydrolysis is the scission of susceptible molecular functional groups by reaction with water. It may be catalysed by acids, bases, salts or enzymes. It is a single-step process in which the rate of chain scission is directly proportional to the rate of initiation of the reaction. A polymer's susceptibility to hydrolysis is the result of its chemical structure, its morphology, its dimensions and the body's environment.

In a commonly used category of hydrolysable polymeric biomaterials, functional groups consist of carbonyls bonded to heterochain elements (O, N, S). Examples include ester amides, urethanes and carbonates and anhydrides. Other polymers containing groups such as ether, acetal, nitrile, phosphonate, sulphonate, sulphonamide or active methylenes hydrolyse under certain conditions. Hydrolytically susceptible groups exhibit differing rates of degradation, which are dependent on the intrinsic properties of the functional group and on other molecular and morphological characteristics. Among carbonyl polymers, anhydrides display the highest hydrolysis rates followed in order by esters and carbonates. Other carbonyl groups such as imide, urethane, amide and urea can demonstrate long-term stability *in vivo* if contained in a hydrophobic backbone or highly crystalline morphological structure.

The rate of hydrolysis tends to increase with a high proportion of hydrolysable groups in the main or side chain, other polar groups which enhance hydrophilicity, low crystallinity, low or negligible cross-link density, a high ratio of exposed surface area to volume and very likely mechanical stress. Porous hydrolysable structures undergo especially rapid property loss because of their large surface area. Factors that tend to suppress hydrolysis rate include hydrophobic moieties (e.g. hydrocarbon or fluorocarbon), cross-linking, high crystallinity due to chain order, thermal annealing or orientation, low stress and compact shape.

FACTORS INFLUENCING HYDROLYSIS RATE

Relative bond stability Hydrolysis rate of a biomaterial is based upon the bond half-life of that material at different physiological pH or temperature. Half-life period is 0.1 h for poly(anhydrides); 4 h for poly(orthoesters); 3.3 years for poly(ester); 83,000 years for poly(amides).

Hydrophobicity Degradation rate generally depends on the hydrophobic status of a polymer. Increase of degradation rate is possible only by decreasing the hydrophobicity of a material, because the material will not allow water penetration, when it has a greater hydrophobicity. Based on this criteria it is possible to determine a material's degradation ability. The cleavage rate of poly(caprolactone) (PCL) is very less compared to poly(D, L-lactide) (PDLLA). Which in turn is less than that of poly (lactide-co-glycolic) acid (PLGA), i.e., **PCL < PDLLA < PLGA.**

Steric effects The hydrolysis rate towards steric effects is mainly based on three factors. They are local structure, glass transition (T_g) and crystallinity. Polylactic acid and polyglycolic acid are the suitable examples for local structure. Polylactic acid degrades slower than polyglycolic acid due to the presence of methyl group in PLA which prevents water penetration, i.e., less water penetration leads to slower degradation of PLA. Chain mobility is commonly exhibited by all rubbery polymers, due to the movement of polymeric repeating units present inside the polymer, and thus they tend to have high T_g (glass transition) values. This condition is favourable for water penetration, so the hydrolysis rate of polymer is also based on the value of T_g. Accommodation of water bodies into the polymeric material depends on the sterioisomeric structure of that material. Generally sterioisomeric nature will easily allow the water to penetrate which leads to degradation, e.g., D-lactide, L-lactide and mesolactide.

Production of autocatalytic products Polymers like polylactic acid (PLA), polyanhydrides and polyesters are suitable examples for self-destroying polymers. PLA is destroyed easily when the inner neutral pH decreases or becomes acidic. In polyanhydrides, the presence of two acid chain ends causes rapid autocatalysis and brings down the pH, making it acidic, which leads to degradation. Similarly, polyesters tend to produce acid chain ends by breakdown process and drop the pH drastically.

Phase separation Phase separation of two blended polymers is easier than that of an individual polymer. For example, in a given material, if there is only lactic acid because of its high crystallinity there will be no degradation. But when it is blended with glycolic acid, its crystallinity is affected by glycolic acid and the rate of degradation increases. Thus changes in the characteristics of any polymeric material will allow water penetration which leads to degradation.

DEGRADABLE MATERIALS FOR BIOLOGICAL RECOGNITION

Generally cells use receptor–receptor/receptor–ligand interactions to guide their functions which includes adhesion, migration, growth and differentiation. The process of differentiation happens by identifying secretion of molecules, binding of molecules and specialized functions. Cells interactions with simple synthetic materials are governed by nonspecific interactions, e.g. surface energies, hydrophobic interactions and charge–charge interactions. But the cells only interact with extracellular matrix.

The extracellular space in the tissues of multicellular animals' are filled with a gel-like material, the extracellular matrix, also called ground substance, which holds the cells together and provides a porous pathway for diffusion of nutrients and oxygen to individual cells. The extracellular matrix is composed of an interlocking meshwork of heteropolysaccharides and fibrous proteins such as collagen, elastin, fibronectin and laminin. Thus the extracellular matrix plays a vital role in providing mechanical support and this gives hint for cell survival/function by showing anchorage-dependent cell growth and differentiation.

Adhesion protein, designed to bind to structural ECM components, present binding sites to receptors. Thus adhesion proteins can present multiple binding sites for different receptors that work in synergy. The interactions of cells with native ECM, by receiving signals from the extracellular environment like growth factors, cytokines, extracellular matrix and chemokines. Cells interact with specific adhesion motifs in adhesion proteins via cell-surface receptors. The microstructure of ECM protein arrangement and its composition can tune adhesion. Adhesion in turn regulates growth, differentiation and migration. The major families of cell-ECM receptors are "Integrins".

Several families of integral proteins in the plasma membrane provide specific points of attachment between cells, or between a cell and ECM proteins. Integrins are heterodimeric proteins (two unlike subunits, α and β) anchored to the plasma membrane by a single hydrophobic transmembrane helix in each subunit. The large extracellular domains of the α and β subunits combine to form a specific binding site for extracellular proteins such as collagen and fibronectin. As there are 18 different α subunits and at least 8 different β subunits, a wide variety of specifications may be generated from various combinations of α and β. One common determinant of integrin binding in several extracellular partners of integrins is the sequence Arg-Gly-Asp (RGD). Figure 2.3 shows RGD-mediated cellular responses.

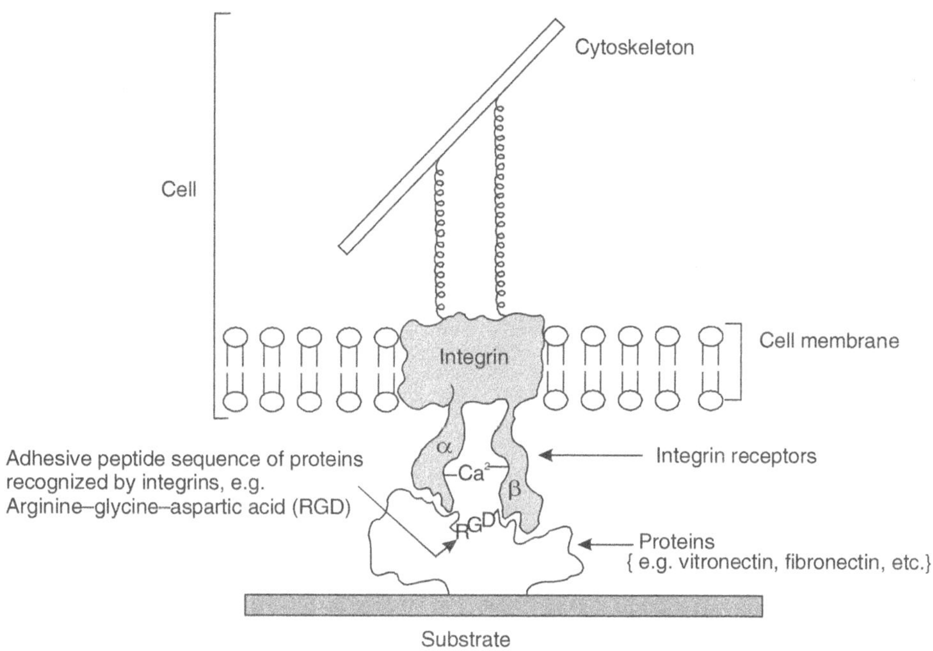

FIGURE 2.3 CELL RECOGNITION OF BIOMATERIAL SURFACES BY PEPTIDE (RGD) COUPLED PROTEIN INTERACTION WITH INTEGRIN RECEPTORS

Integrins are not merely adhesives, they serve as receptors and signal transducers, conveying information across the plasma membrane in both directions. Integrins regulate many processes, including platelet aggregation at the site of wound, tissue repair, the activity of immune cells and the invasion of tissue by a tumour. Integrin clustering drives actin filament assembly and can signal through multiple biochemical pathways, some of which synergies with growth factors to the cell "where it is". Biochemical signal triggers affinity change integrins and there are 20 different integrins, many different pairs possible with different ligand specificities, thus cell-specific adhesion can be modulated by ECM.

How do we make materials that can interact in an engineered way with their biological environment? The answer is simply by incorporation of peptides into the synthetic polymers. Peptides have been introduced in the polymers to provide scaffolds that appear less foreign and have some engineered response from cells. Peptide sequences conjugated to synthetic polymers have been used as signals for adhesion (fragments from ECM adhesion proteins) and to support transformation of synthetic scaffolds into de novo natural matrix, to support cell migration through solid scaffolds. The reason for using peptides instead of proteins is that they are fragile in nature, not soluble in organic solvents (but peptides often are), costly, and the immunogenicity of peptides is less than complete protein sequences (to reduce likelihood of provoking inflammatory response to devices).

BIOMATERIALS IN TISSUE ENGINEERING

Tissue engineering generally defined as "the process of creating living, physiological, three-dimensional tissues and organs utilizing specific combinations of cells, cell scaffolds, and cell signals, both chemical and mechanical." As the field of tissue engineering has developed, novel biomaterials have been created to facilitate these approaches. Currently there are three ways in which biomaterials have been utilized in the field of tissue engineering:

1. To induce cellular migration or tissue regeneration
2. To encapsulate cells and act as an immuno-isolation barrier
3. To provide a matrix to support cell growth and cell organization.

The first approach involves the use of bioactive biomaterials to facilitate local tissue repair. Such materials have been developed in several forms (e.g. powders, solutions, or doped microparticles) and have been designed to release a variety of chemicals, proteins, and/or growth factors in a controlled fashion by diffusion or network breakdown. These agents trigger a local cellular response leading to tissue repair and regeneration. The infusion of bone morphogenic protein into orthopaedic implants as well as the use of glycosaminoglycan and collagen constructs to act as an artificial skin substitute are current examples of biomaterials used in tissue engineering.

In the second strategy, a polymer is used as an immuno-isolation barrier for encapsulated cells. For example, hollow fiber membranes enclosing hepatocytes have been used to construct a bioartificial liver for the treatment of liver failure. Similarly, microcapsules that store and protect cellular transplants also have been developed. These microcapsules protect the transplanted cells from immune cell penetration while allowing the passage of medium- and small-sized substances, such as oxygen, nutrients, and wastes.

The third and most widespread use of biomaterials in tissue engineering involves the use of tissue scaffolds to direct the three-dimensional organization of cells *in vitro* and *in vivo*. Scaffolds are porous structures created from natural or synthetic polymers. Scaffolds can be constructed in a variety of forms including spongelike sheets and fabrics, gels, and highly complex three-dimensional structures with detailed pores and channels. Similar to the biomaterials of the 1950s and 1960s, the early tissue engineering scaffolds were adapted from other uses in surgery. These early scaffolds, often made of surgical fabrics and mesh, provided a three-dimensional biocompatible structure onto which cells could attach. However, they did not specifically interact with cells in a controlled fashion. More recently, scaffolds are being created with embedded growth factors or cellular ligands, allowing them to influence signalling pathways necessary for cell migration and proliferation. In this manner, these new scaffolds can be thought of as synthesized extracellular matrix, providing tissues with the appropriate architecture and signalling pathways to influence key cell functions. Currently, scaffolds have been used to engineer a variety of tissues including vascular tissue, skin, bone, and cartilage.

MICRO/NANOTECHNOLOGY AND BIOMATERIALS

In recent years, significant interest has developed surrounding the utilization of micro- and nanofabrication techniques for the construction of novel biomaterials and implants. Developed from advances in the fields of computer science, engineering, physics, and biology, these techniques have found widespread application in an array of industries including computing, consumer electronics, manufacturing, and biotechnology. While the application of these techniques toward biomaterials is relatively new, they promise to enable the creation of smaller, more active, and more dynamic implants and devices.

Microfabrication and Microtechnology

Microfabrication is the process of constructing materials and devices with dimensions in the micrometre to millimetre range. Developed and refined for the processing of integrated circuits, microfabrication techniques have been utilized since the 1950s for the production of semiconductor-based microelectronics. In general, microfabrication techniques utilize a "top down" approach for creating structures, where one takes a substrate and builds a device out of the bulk material (bulk micromachining) or on its surface (surface micromachining). These techniques include unit process steps such as thin-film growth/deposition, photolithography, etching, and bonding and will be covered in detail in later chapters. Over the last few decades, microfabricated devices containing electrical and mechanical components, also known as microelectromechanical systems (MEMS), have made their way into a multitude of products used in daily life. For example, the air bag system in an automobile is likely to include a MEMS accelerometer to help the system detect the proper time to fire. Similarly, in the health care arena, MEMS technology can be found in blood pressure sensors, blood chemistry analysis systems, DNA array systems, and home telemetry monitoring systems.

In the field of biomaterials and implants, MEMS technology is being applied toward the development of implantable drug delivery systems. Such systems utilize micropumps that deliver regulated amounts of stored medication in a controlled fashion. The use of such systems for the delivery of insulin for the treatment of diabetes is currently being investigated.

Nanofabrication and Nanotechnology

Nanotechnology has been defined as "research and technology development at the atomic, molecular, and macromolecular levels in the length and scale of approximately 1–100 nanometre range, to provide a fundamental understanding of phenomena and materials at the nanoscale and to create and use structures, devices, and systems that have novel properties and functions because of their small and/or intermediate size." On a simpler level, this definition can be broken down into two specific and important characteristics that distinguish nanotechnology and nanostructures from their micro- and macro-counterparts. First, nanotechnology refers to the creation and manipulation of structures at the nanoscale range (1–100 nm). Second, nanotechnology is concerned with the characterization and application of the unique physical, chemical, and biological properties that nanoscale structures display because of their size. New behaviour at the nanoscale range is not necessarily predictable and may be completely different from that of the same material in its bulk, macroscopic form. The understanding and utilization of these properties is a fundamental goal of nanoscience and nanotechnology.

Nanofabrication refers to the processes and methods employed in the creation of nanoscale materials and structures. In contrast to the microfabrication techniques developed for the semiconductor and MEMS industries, nanofabrication techniques utilize complementary "top down" and "bottom up" fabrication methods. Building on advances in microfabrication, nanoscience researchers have utilized newer "top down" fabrication methods such as electron beam lithography to yield near-atomic scale precision. Alternatively, the "bottom up" creation of nanostructures is being developed where fabrication starts at the molecular level. Structures are "self-assembled" by taking advantage of the atomic and molecular properties of nanoscale materials. In this manner, nanofabrication can be thought of as being inspired by nature, as biological structures are typically assembled and rearranged at the nanoscale range using molecular interactions such as van der Waals forces, hydrogen bonds, and electrostatic dipoles.

Today, nanotechnology is an exploding field. Since the year 2000, more than 35 countries have developed programs in nanotechnology. Furthermore, worldwide government funding of nanotechnology has increased approximately five-fold since 1997, exceeding 2 billion dollars in 2002. In the United States alone, the National Nanotechnology Initiative, established by President Bill Clinton in 2000, has grown rapidly in both scope and support, with a 2002 federal budget award of approximately $697 million. With this intense interest and development, nanotechnology and nanofabrication have already played a significant role in the development of new medical devices and materials. Today, pharmaceutical preparations are being encapsulated in a variety of nanostructures to enhance effectiveness and decrease side effects. Similarly, nanoparticle-based scaffolds that facilitate bone growth have already received FDA approval for use in orthopaedic surgery. As we look toward the future, nanotechnology will undoubtedly play a significant role in the development and use of future generations of biomaterials.

CONCEPT CHECK

1. Why and how Biodegradable Polymers are important in human welfare?
2. Give examples for mechanical implants.
3. What is the contribution of Göpferich theory?

MIND TWISTER

1. In Poly(alkyl cyanoacrylates), will the presence of cyano group have any harmful effect on the body?

REFERENCES

Bajpai, P.K. and Billote W.G. (1995). Ceramic biomaterials. In: *The Biomedical Engineering Handbook*. Bronzino, J.D. (ed.). CRC Press, Boca Raton, FL.

Catledge, S.A., Fries, M.D., Vohra Y.K. *et al.* (2002). Nanostructured ceramics for biomedical implants. *J. Nanosci. Nanotechnol.* 2(3–4), 293–312.

Davis, J.R. (2003). Overview of biomaterials and their use in medical devices. In: *Handbook of Materials for Medical Devices*. Davis, J.R. (ed.). ASM International. Materials Park, OH. pp. 1–13.

Edlund, U. and Albertsson, A.C. (2002). Degradable polymer microspheres for controlled drug delivery. *Adv. in Polymer Sci.* 157, 67–112.

Griffith, L.G. and Naughton, G. (2002). Tissue engineering: Current challenges and expanding opportunities. *Science*. 295, 1009–1014.

Hench, L.L. and Polak, J.M. (2002). Third-generation biomedical materials. *Science*. 295(5557), 1014–1017.

Machado, L.G. and Savi, M.A. (2003). Medical applications of shape memory alloys. *Braz. J. Med.Biol. Res.* 36(6), 683–691.

Maheshwari, G., Brown, G., Lauffenburger, D.A., Wells, A. and Griffith, L.G. (2000). Cell adhesion and motility depend on nanoscale RGD clustering. *J. Cell Sci.* 113, 1677–86.

Park, J.B. and Lakes, R.S. (1992). *Biomaterials: An Introduction.* Plenum Press, New York.

Piskin, E. (2002). Biodegradable polymeric matrices for bioartificial implants. *Int. J. Artif. Organs.* 25(5), 434–440.

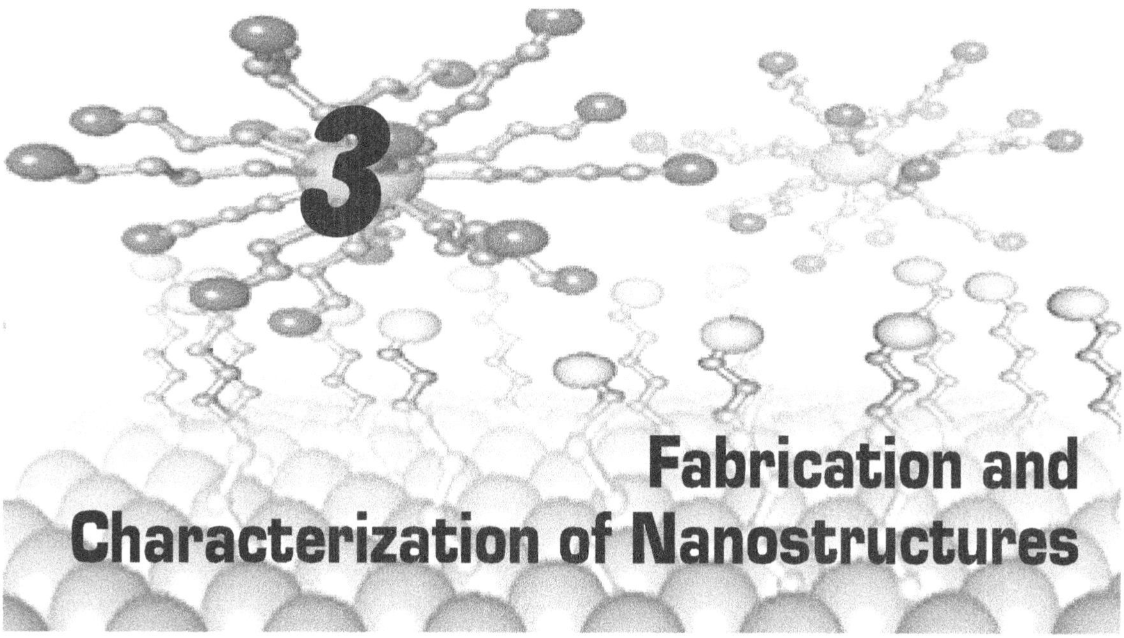

3 Fabrication and Characterization of Nanostructures

INTRODUCTION

Nanostructured materials, particularly those derived from nanoparticles, have evolved as a separate class of materials over the past decade. The most remarkable feature has been the way in which completely disparate disciplines have come together with nanomaterials as the theme. The breadth of the field is enormous, ranging from the use of nanoparticles of zinc oxide in hygiene products, such as diapers, to altering the characteristics of solid rocket propellants by the addition of nanoparticle fillers. The enthusiasm is justified for the most part, as the fundamental materials' properties appear to be different at the nanoscale. There is also ample evidence that nanoparticles display characteristics that are distinctly different from their microcrystalline counterparts. The field has matured so rapidly and so fast that it is probably hard to find a segment of any technical subject where the implications of nanomaterials have not been explored at least to a preliminary extent. Studies are being conducted on the potential use of nanomaterials in diverse applications, including hydrogen storage, ion-sensing and gas sensing, surface-modified nanoparticles for enhanced oil recovery, adsorption of chemical and biological agents on to nanoparticles, active electrode materials for lithium-ion batteries, light-emitting devices and dental compositions, to name a few.

The purpose of this chapter is to examine the progress to date in the science and technology of the production of ultrafine particles, which form the building blocks for nanostructured materials. While high surface area inorganic materials of a few compositions, such as supported metal catalysts, carbon black, and nanoparticulate silica, have been in use for several decades, it was not until the 1970s and 1980s that new techniques were developed in a deliberate attempt to synthesize tailored nanoparticles of different compositions. However, these efforts were few and far apart. Further, the implications of materials at such a fine scale were at best poorly appreciated. Some 10 to 15 years later, in the late 1980s and early 1990s, several new nanoparticle synthesis techniques began to be developed, most of them at university and national research laboratories. With the availability of nanoparticles, albeit in relatively small quantities, applications began to be developed, which in turn provided the impetus to scale the production processes to commercially accepted levels, and at the same time, develop lower cost approaches to nanoparticles synthesis. Sometime thereafter, the emphasis shifted to synthesis of "designer" nanoparticles, often with a complex structure as well. These were aimed at specific applications, where either the nanoparticles when added to a matrix would provide a desired functionality, or the nanoparticles themselves were processed into a film or a coating. In contrast, despite a massive global effort, processing nanoparticles into large bulk nanostructured objects has remained a laboratory curiosity. This chapter is an attempt to outline the progression of the field of nanoparticulate materials from both a theoretical understanding and followed by fundamental for research.

NANOPARTICLE SYNTHESIS TECHNIQUES: METHODOLOGY AND CLASSIFICATION

All particle synthesis techniques fall into one of the three categories: vapour-phase, solution precipitation, and solid-state processes. Although vapour-phase processes have been in vogue during the early days of nanoparticle development, solid-state process is the most widely used in the industry for production of micron-sized particles, mainly due to cost considerations. The following is a description of the solid-state nanoparticle synthesis process.

Solid-State Synthesis of Nanoparticles

Solid-state synthesis generally involves a heat treatment step in order to achieve the desired crystal structure, which is followed by media milling. Nanoparticles as small as 30 nm can be produced using milling media. The scientific community has not shown much enthusiasm for mechanical attrition processes for nanoparticle synthesis, perhaps due to issues pertaining to impurity pick up, lack of control on the particle size distribution, and inability to tailor precisely the shape and size of particles in the 10–30 nm range, as well as the surface characteristics. Nonetheless, in several instances a modified version of mechanical attrition has been used to synthesize oxide nanoparticles. In the mid-1990s, Advanced Powder Technology in Australia pioneered a solid-state process with a post-milling operation. Dry milling was used to induce chemical reactions through ball-powder collisions that resulted in forming nanoparticles within a salt matrix. Particle agglomeration was minimized by the salt matrix, which then was removed by a simple washing procedure. One of the major products was cerium oxide, which generally has been an expensive material and scarcely available in a nanopowder form. There continues to be reports on the so-called mechanochemical processing of nanoparticles, even though in various different forms.

However, mechanical milling is a versatile technique to produce metallic nanocrystalline micropowders, as opposed to high surface area nanoparticles. One of the earliest efforts on nanocyrstalline metals, was the ball-milling of aluminium and its alloys in liquid-nitrogen atmosphere. In addition to refining the grain size, the milling process introduced ultrafine dispersion, which improved the high temperature creep resistance. In conventional high-energy ball-milling, the creation and self-organization of dislocations to high-angle grain boundaries within the powder particles during the milling process leads to a reduction in the grain size by a factor of about 10^4.

Vapour-Phase Synthesis of Nanoparticles

Gas condensation, as a technique for producing nanoparticles, refers to the formation of nanoparticles in the gas phase, i.e., condensing atoms and molecules in the vapour phase. For example, Cabot Corporation in the United States and Degussa in Germany, have been using atmospheric flame reactors for decades to produce megatons of such diverse nanoparticles as carbon black (used in tires and inks), silicon dioxide (used in myriad applications including additives in coffee creamers and polymers), and titanium dioxide (used in scores of applications including UV-protecting gels). The generic process involves hydrolysis of gaseous metallic chlorides under the influence of water, which develops during the oxyhydrogen reaction, which in turn, leads to a high-temperature reaction zone. The reaction products include the oxide powder and hydrochloric acid, which are recycled. The powders have relatively high surface area (e.g. 50 m^2/g for TiO_2; primary particle size: 21 nm) and disperse to the extent necessitated by the application.

Inert gas condensation of nanoparticles The general perception was that the particle size, shape, and extent of aggregation could be considerably controlled if gas condensation processes could be carried out either in a low-pressure environment, or the nanoparticles were quenched rapidly as soon as they were formed. Initially metal nanoparticles were synthesized by the inert gas condensation (IGC) process. The nanoparticles, some of them with a mean size of 10 nm and smaller were formed when metal atoms effusing from a thermal source rapidly lost their energy by collisions with gas atoms. Figure 3.1 shows a schematic of the set-up to produce metal nanoparticles. A number of metal nanopowders, including Al, Co, Cr, Cu, Fe, Ga, Mg, and Ni, were synthesized by this technique. A large glass cylinder (diameter, 0.34 m; height, 0.45 m) fitted to water-cooled stainless steel end plates was evacuated to a pressure of approximately 2×10^{-6} torr by an oil diffusion pump. An alumina crucible placed on a stand-off was slowly heated via radiation from a graphite heater element. After appropriate outgassing, the pump line was closed and a reduced atmosphere of an inert gas, usually 0.5 to 4 torr of high-purity argon, was introduced into the cylinder. The crucible was now heated rapidly

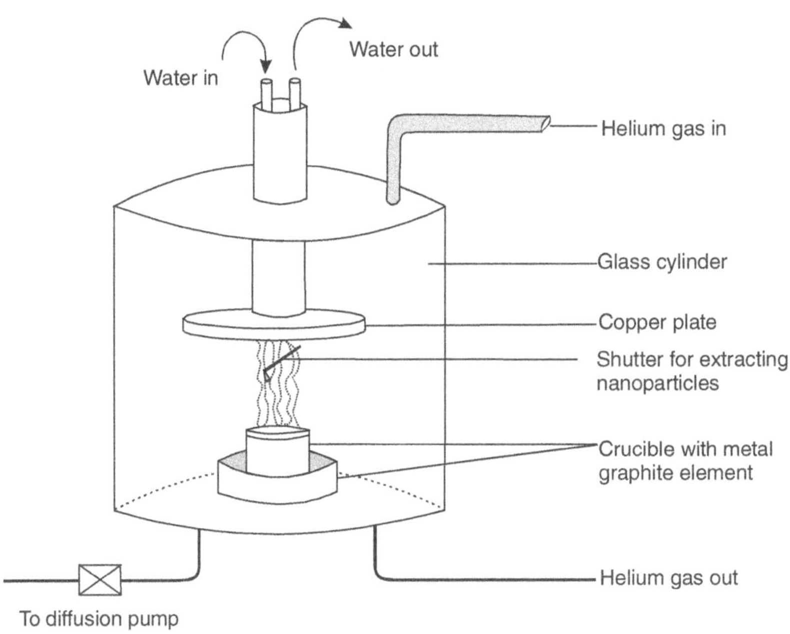

FIGURE 3.1 SCHEMATIC SET-UP OF INERT GAS CONDENSATION

under quasi-equilibrium conditions (constant temperature and inert-gas pressure). The nanoparticles, which nucleate and grow in the gas phase, were collected on a water-cooled copper surface. The production rate was about 1 g per run.

The short collision mean free path, e.g. 10^{-7} m for aluminium at 1 torr of argon, resulted in efficient cooling, producing a high supersaturation of metal vapour, leading to homogeneous nucleation. It was shown that the dominant mechanism of particle growth was by coalescence of clusters into nanoparticles, which resulted in the formation of nanoparticles with a log normal size distribution. Some researchers introduced a modification to the process by carrying it out in an ultrahigh vacuum chamber (backfilled with 1 to 10 torr of inert gas) and condensing the nanoparticles on the so-called "coldfinger," which was a liquid-nitrogen-filled rotating cylinder. Figure 3.2 shows a schematic set-up of the process; nanoparticles develop in a thermalizing zone just above the evaporative source due to interactions between the hot vapour species and the much colder inert gas atoms in the chamber. The process is versatile as both metals and oxides could be synthesized, the latter being enabled by the introduction of oxygen. As such, the "cold-finger" was convenient to deposit the nanoparticles, scrape off and pack into a die, all under quasi ultrahigh vacuum conditions.

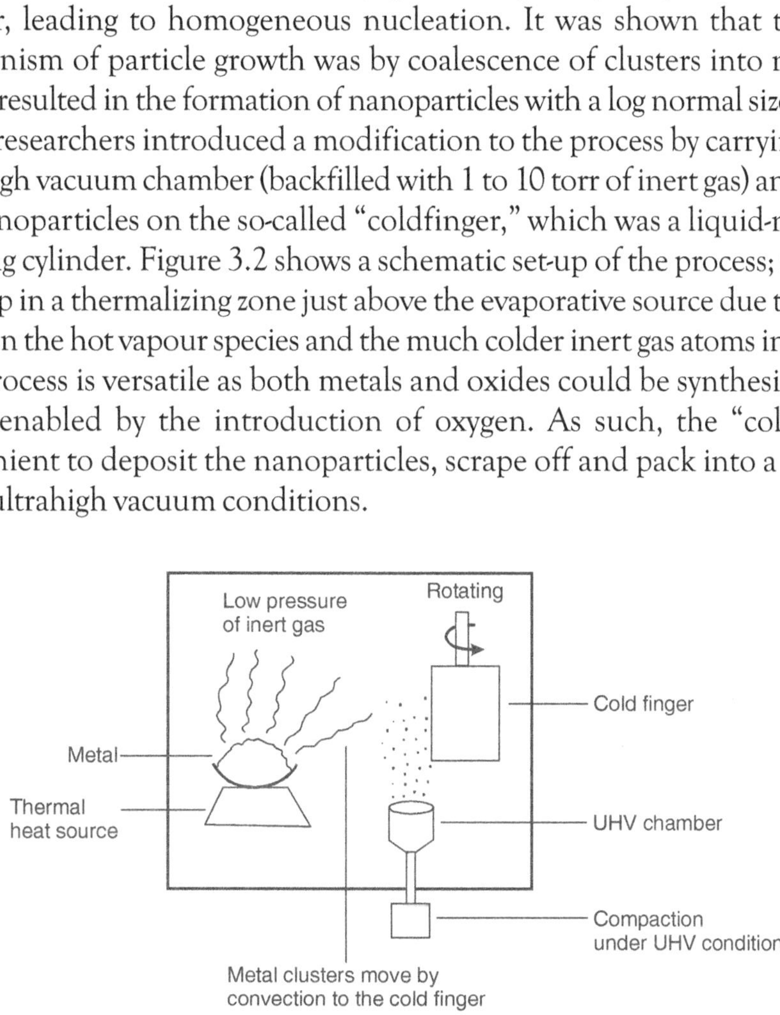

FIGURE 3.2 SCHEMATIC SET-UP OF "COLD FINGER" BY GLEITER

Plasma-based synthesis of nanoparticles A thermal plasma (i.e., ionized gases), as a heat source for melting materials had been gaining prominence for quite some time in the materials community. In fact, plasma spraying of materials on to substrates to form protective coatings has been a well-established industrial practice for decades. So it was natural for researchers to begin using a thermal plasma as a heat source for evaporating materials, both metals and ceramics. A piece of metal was mounted on a water-cooled copper hearth and heated by a plasma jet flame. The gas atmosphere was helium (a few hundred torr), mixed with about 15% hydrogen. The evolving smoke flowed into a cold cone on to which the ultrafine particles deposited. They were collected in an ampoule, which could be sealed off without exposing the collected sample to air. Using a 10 kW plasma gun, ultrafine particles of Al, Co, Cu, Fe, Ti, and Ta were produced with the scaled equipment. The mean diameter was ~ 20 nm and the production rate was as high as 50 g/h for some metals. However, problems, such as not being able to focus the plasma at a pressure lower than 200 torr and deterioration of the plasma in long runs, led to curtailment of further development of the process.

Some nanotechnologies developed a direct current arc plasma method to produce metal nanoparticles. Essentially, a metal block was placed on a water-cooled copper anode plate or a graphite anode crucible. The equipment was evacuated to about 100 Pa and then backfilled with a hydrogen–argon mixture gas at 0.1 MPa pressure, at which point the sample was melted by the arc. Metal vapours were formed and condensed in the gas phase, and subsequently removed by circular gas flow. Hydrogen gas, free of any entrained powder, was reintroduced into the generation chamber by a gas circulation pump. It was found that the presence of hydrogen increased the rate of formation of nanoparticles by a factor of 10 to 10,000 compared to conventional evaporation. The enhanced evaporation mechanism was explained as follows: in arc melting under a hydrogen atmosphere, the molten metal comes into contact with both atomic as well as molecular hydrogen. The former is substantially more soluble in the metal and reaches a supersaturated state rapidly. A supersaturated hydrogen beyond the arc evolves into non-arc gas phase. The extensive dissolution and subsequent evolution of hydrogen gas from the molten metal leads to enhanced evaporation of metal.

A number of variations of the plasma approach began to emerge in development work across the globe, each improving upon the previous work. Some researchers argued that the high temperatures in a plasma lead to cold boundary layers and nonuniformities in processing conditions, especially during condensation. They developed a modified process wherein nonuniformities were minimized by expanding the plasma containing the vapour-phase precursors through a subsonic nozzle with a hot ceramic wall. This arrangement approached a configuration of one-dimensional flow with one-dimensional temperature gradients in the direction of the flow in the nozzle, leading to high uniformity of the quench rate. Furthermore, the nozzle provided much higher quench rates than would have been obtainable otherwise.

The reactor consisted of a water-cooled chamber with an assembly consisting of a plasma torch, a reactant injection system, and a converging nozzle mounted on the top flange of the chamber. The torch was a Miller SG-1B plasma gun with a special tungsten-lined nozzle for argon–hydrogen operation. The 25-mm-long injection section was immediately downstream of the anode, and consisted of a water-cooled nickel ring with a ceramic liner and holes at two axial locations, which are connected to the reactant supplies. Precursor vapours were introduced through heated lines, and immediately following the injection section is the 50-mm-long converging nozzle held in place by a water-cooled nickel holder. The nozzle and the liner for the injection ring were made of one piece of boron nitride. The plasma system were shown in Figure 3.3.

Flame-based synthesis of nanoparticles The use of a hydrocarbon (or hydrogen)–oxygen flame to pyrolyse chemical precursor species and produce nanoparticles is attractive in principle due to the fact that flame processes are already in use on a commercial scale. Over the past decade and a half, research has been directed predominantly toward introducing uniformity and control over the pyrolysis process in a flame, with the anticipation of forming nanoparticles with a narrow size distribution and minimal aggregation. This included developing flames with a flat geometry, as opposed to the traditional Bunsen burner conical flames. There were predominantly two variations to the theme of nanoparticle synthesis using combustion flames. On the one hand, researchers worked extensively on atmospheric flames, including studies on the effect of an electric field on the flame itself as well as on the nanoparticles. This led to an increased understanding of cluster formation and particle growth in a flame. While the traditional flame processes involved mixing reactants along with the combustibles, a variation of the process was the counter-current flow scheme, in which reactants were fed independently through a separate tube into the flame. A further variation on the theme of separating the reactant stream from the fuel/oxidizer stream was a multi-element diffusion flame burner. Wooldridge and co-workers have described one such design, where the flat flame diffusion burner consists of an array of hypodermic needles set in a honeycomb matrix: the fuel is mixed with the precursor reactant and flowed through individually sealed tubes that are less than a millimetre in diameter, and the oxidizer flows through the surrounding channels in the honeycomb. The processes described above utilized precursor vapours, and so were restricted to oxide ceramics that could be derived from metalorganic or organometallic precursors with ambient pressure boiling points of ~200°C or lower. As such, acetates and nitrates were not utilized. The atomized nonvolatile precursors dissolved in a solvent and directed them through a combustion flame. Powders with high surface area were formed as a result of rapid pyrolysis. The range of composition of nanopowders could now be expanded to multicomponent oxide materials. It should be noted that the mechanism of nanoparticle formation is unlike that of vapour condensation when nonvolatile species are pyrolysed in the flame.

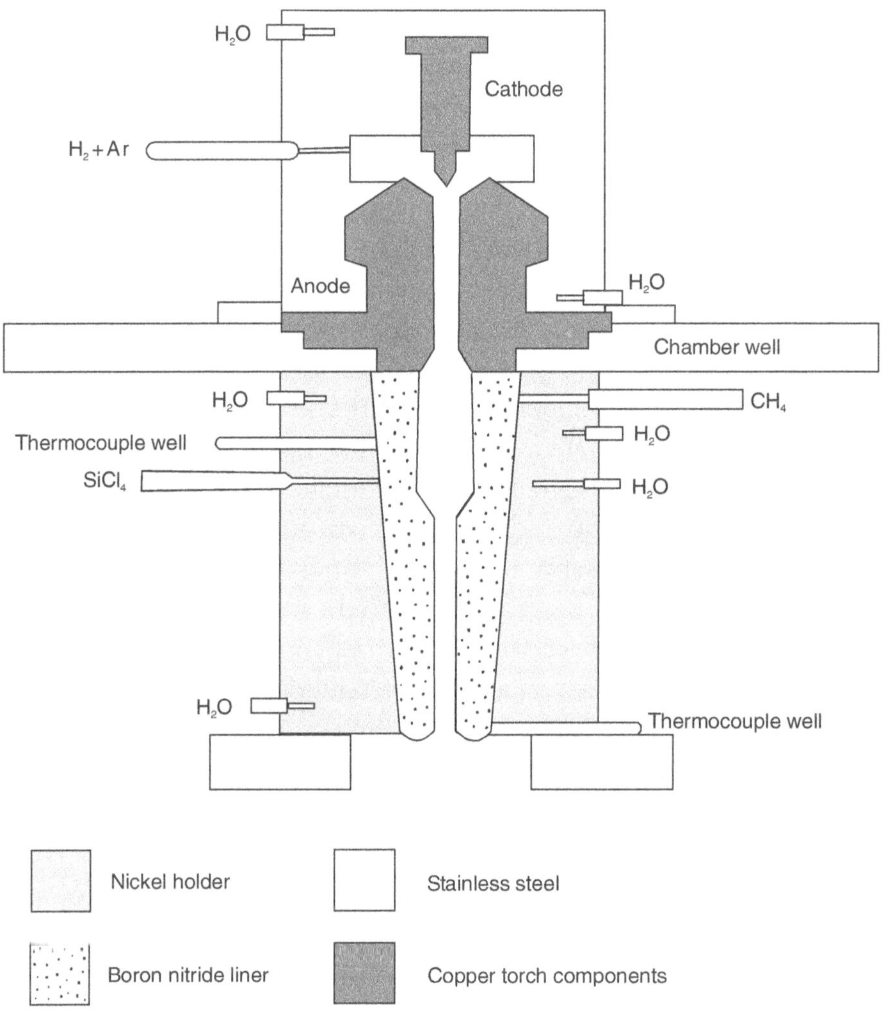

FIGURE 3.3 PLASMA PROCESS WITH EXPANDING NOZZLE TO PRODUCE NANOPARTICLES

A majority of the vapour-phase processes have until now been directed toward the synthesis of oxide ceramics, since they form the bulk of the ceramics industry. Figure 3.4 shows the nanoparticle synthesis process by vapour condensation method. Hot-wall reactors have also been employed to synthesize nanostructured SiC and Si_3N_4 powders. The relatively slow kinetics of nitridation and carburization reactions has been the fundamental issue with the ability to synthesize nitride and carbide nanoparticles in the gas phase. The short residence at high temperatures in any vapour-phase process is just not enough to complete the reaction. By evaporating nitriding aluminium nanoparticles *in situ* in a forced-flow reactor with microwave plasma downstream, which dissociates nitrogen molecules in the gas stream and reheats the particles to promote complete conversion of Al to AlN.

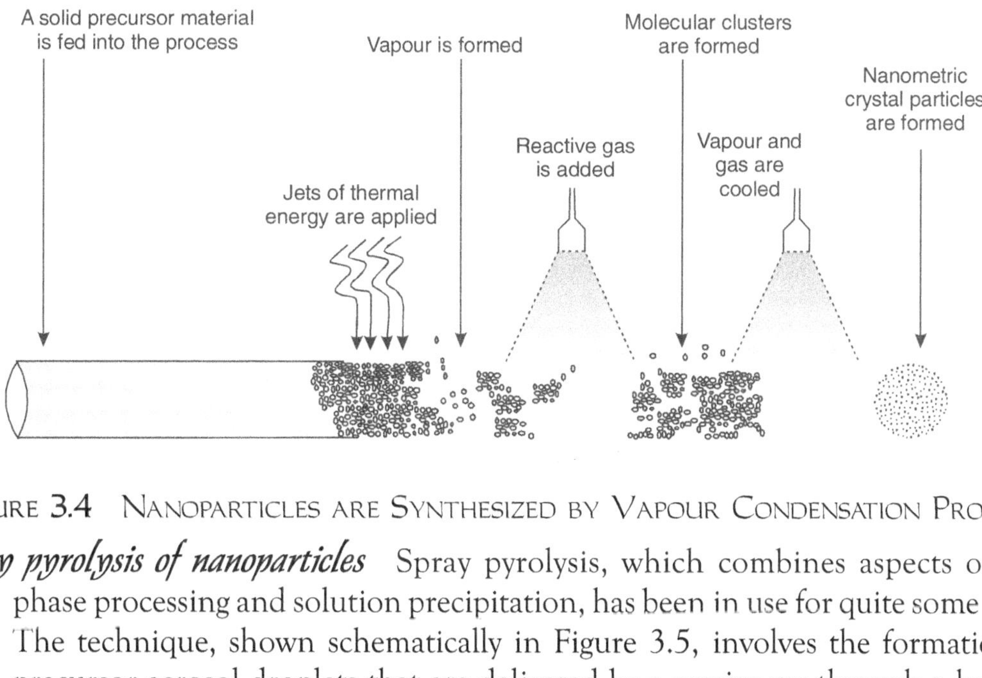

FIGURE 3.4 NANOPARTICLES ARE SYNTHESIZED BY VAPOUR CONDENSATION PROCESS

Spray pyrolysis of nanoparticles Spray pyrolysis, which combines aspects of gas-phase processing and solution precipitation, has been in use for quite some time. The technique, shown schematically in Figure 3.5, involves the formation of precursor aerosol droplets that are delivered by a carrier gas through a heating zone. Precursor solutions of metal nitrates, metal chlorides, and metal acetates are atomized into fine droplets and sprayed into the thermal zone. Inside the heating zone, the solvent evaporates and reactions occur within each particle to form a product particle. Spherical, dense particles in the 100–1000 nm range can easily be formed in large volume by this method. The principal advantage of the spray pyrolysis method is the ability to form multicomponent nanoparticles as solutions of different metal salts can be mixed and aerosolized into the reaction zone.

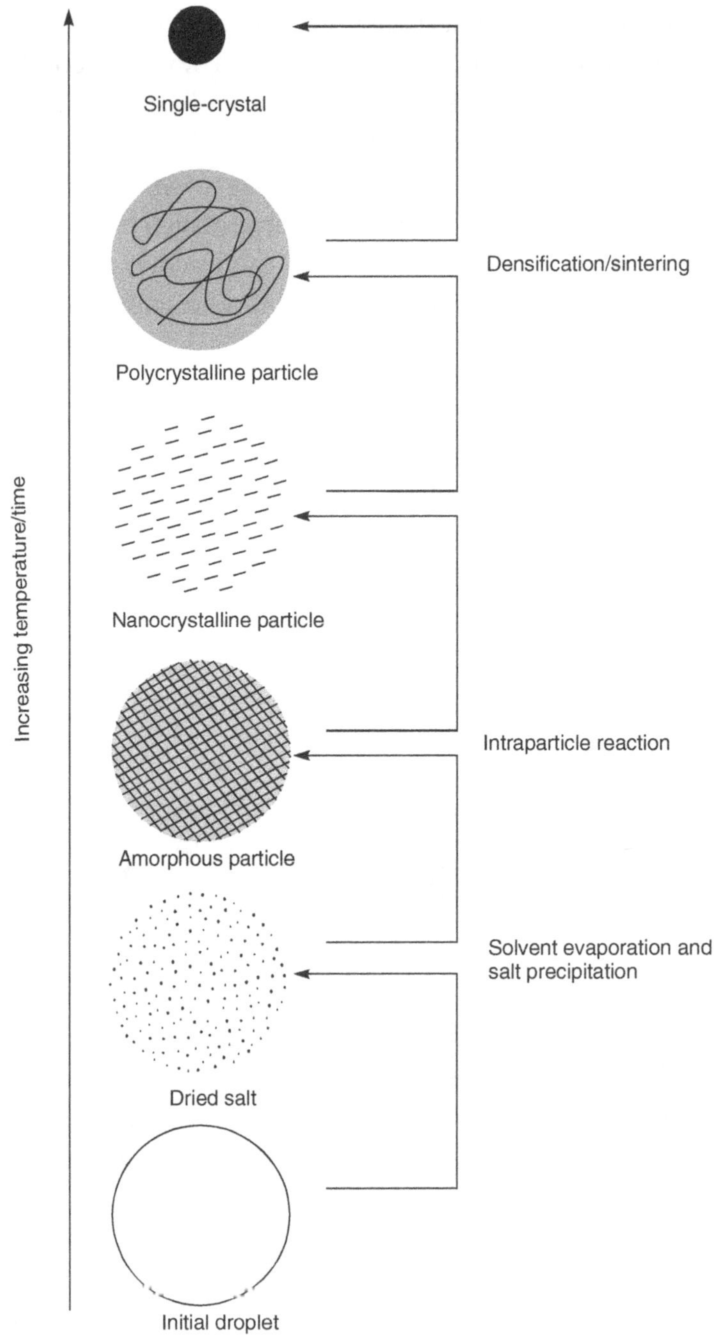

FIGURE 3.5 EVENTS DURING THE SPRAY PYROLYSIS PROCESS

Over the years, much emphasis has been given to being able to reduce the precursor droplet size during spray pyrolysis as this in turn would reduce the particle size. The limit on the particle size was pushed further by using precursor drops that were 6–9 μm in diameter. Uniformly sized dense and spherical particles of ~ 150 nm, were obtained by reducing the precursor concentration about an order of magnitude. In a modified version of the spray pyrolysis method, researchers synthesized SiO_2-encapsulated Pd nanoparticles. These composite nanoparticles were formed from a Pd-nitrate solution containing ultrafine SiO_2 particles by ultrasonic spray pyrolysis. A precursor particle, formed below 700°C in the drying stage, was composed of a homogeneous mixture of nanoparticles of SiO_2 and $PdO{\cdot}H_2O$. When PdO was decomposed above 700°C, metallic Pd nanoparticles were formed in the SiO_2 matrix. Because of the high surface free energy, the Pd particles coalesced and condensed in the interior of the composite particle. As a result of the relocation within the composite particle, SiO_2 was forced out of the particle toward the surface, and an SiO_2-encapsulated Pd particle was formed.

SOLUTION PROCESSING OF NANOPARTICLES

Precipitating clusters of inorganic compounds from a solution of chemical compounds has been an attractive proposition for researchers, primarily because of the simplicity with which experiments can be conducted in a laboratory. This is especially true if the goal is to just have a nanocrystalline powder, instead of a "dispersible" nanoparticulate powder. For example, high ionic conductivity nanocrystalline CeO_2 can be synthesized with consummate ease by precipitating Ce and Gd nitrates in dilute ammonia solution at room temperature. A major advantage of solution processing is the ability to form encapsulated nanoparticles, specifically with an organic molecule, for providing functionality to the nanoparticles, improving their stability in a medium, or for controlling their shape and size. Solution processing can be classified into five major categories: 1) sol–gel processing, 2) precipitation method, 3) water–oil microemulsions (reverse micelle) method, 4) polyol method, and 5) hydrothermal synthesis.

SOL–GEL PROCESSING

Sol–gel technique is one of the most popular solutions processing method for producing metal oxide nanoparticles. Over the years, solution precipitation and sol–gel processing have come to be used interchangeably, mostly by people on the fringes of the technical community. In sol–gel processing, a reactive metal precursor, such as metal alkoxide, is hydrolysed with water, and the hydrolysed species are allowed to condense with each other to form precipitates of metal oxide nanoparticles. The precipitate is subsequently washed and dried, which is then calcined at an elevated temperature to form crystalline metal oxide nanoparticles. The hydrolysis of metal alkoxides involves nucleophilic reaction with water, which is as follows:

$$M(OR)_y + xH_2O \rightarrow M(OR)_{y-x}(OH) + xROH$$

Condensation occurs when either hydrolysed species react with each other and release a water molecule, or a hydrolysed species reacts with an unhydrolysed species and releases an alcohol molecule. The rates at which hydrolysis and condensation reactions take place are important parameters that affect the properties of the final product. For example, slower and more controlled hydrolysis typically leads to smaller particles, and base-catalysed condensation reactions form denser particles.

Solution Precipitation

In the precipitation method, an inorganic metal salt (e.g. chloride, nitrate, acetate, or oxychloride) is dissolved in water. Metal cations in water exist in the form of metal hydrate species, such as $Al(H_2O)^{3+}$ and $Fe(H_2O)_6{}^{3+}$. These species are hydrolysed by adding a base solution, such as NaOH or NH_4OH. The hydrolysed species condense with each other to form either a metal hydroxide or hydrous metal oxide precipitate on increasing the concentration of OH^- ions in the solution. The precipitate is then washed, filtered, and dried. The dried powder is subsequently calcined to obtain the final crystalline metal oxide phase. The major advantage of this process is that it is relatively economical and is used to synthesize a wide range of single- and multicomponent oxide nanopowders. Additionally, nanocomposites of metal oxides are also produced by co-precipitation of corresponding metal hydroxides. One of the major drawbacks of the process as described above is the inability to control the size of particles and their subsequent aggregation.

In the recent past, significant efforts have been made to control particle characteristics, such as surface area and aggregate size, by precipitating in the presence of a surfactant or an organic molecule. Thus mesoporous, high surface area zirconium oxide is synthesized by incorporating cationic quaternary ammonium surfactants in the hydrous oxide and subsequent calcination of the inorganic/organic intermediate. Surfactants were incorporated by cation exchange at a pH that was above the isoelectric point of the hydrous oxide. The pore size distribution and surface area of zirconium oxide nanopowders were modified by changing the length of the hydrophobic chain from C_8 to C_{18}. Changing the precipitating agent from ammonium hydroxide to tetraalkylammonium hydroxide results in a decrease in the primary particle size of yttrium oxide. The reason for the decrease in the particle size was attributed to the higher pH that can be achieved with tetraalkylammonium hydroxide in comparison to ammonium hydroxide and the steric effect of tetraalkylammonium cation to reduce the diffusion of soluble precursors to the particle surface. The key to the formation of nanoparticles was an intermediate peptization step, wherein the particle size of the precipitate is reduced by adjusting the pH to acidic values.

Water–Oil Microemulsion (Reverse Micelle) Method

Uniform and size-controlled nanoparticles of metal, semiconductor, and metal oxides can be produced by the water-in-oil (W/O) microemulsion (also called reverse micelle) method. In a W/O microemulsion, nanosized water droplets, stabilized by a surfactant, are dispersed in an oil phase. A schematic of a W/O microemulsion is shown in Figure 3.6. Nanosized water droplets act as a microreactor, wherein particle formation occurs and helps to control the size of nanoparticles. A unique feature of the reverse micelle process is that the particles are generally nanosized and monodisperse. This is because the surfactant molecules that stabilized the water droplets also adsorb on the surface of the nanoparticles, once the particle size approaches that of the water droplet. A common way to practice the reverse micelle technique is by mixing two microemulsions that carry appropriate reactants. Water droplets of two microemulsions are allowed to collide with each other and the particle formation reaction takes place inside the water droplet. Nanoparticle synthesis inside reverse micelles is accomplished by one of the two different chemical reactions: (i) hydrolysis of metal alkoxides or precipitation of metal salts with a base, in case of metal oxide nanoparticles, and (ii) reduction of metal salts with a reducing agent, such as $NaBH_4$, in case of metal nanoparticles. Particles are either filtered or centrifuged and then washed with acetone and water to remove any residual oil and surfactant molecules adsorbed on the surface of nanoparticles. Subsequently, the powders are calcined to form the final product. One of the issues in this process is being able to remove efficiently the nanoparticles from the organic phase, and simple washing is not sufficient. One way of removing metal and semiconductor nanoparticles from the oil phase is to immobilize them on stable supports.

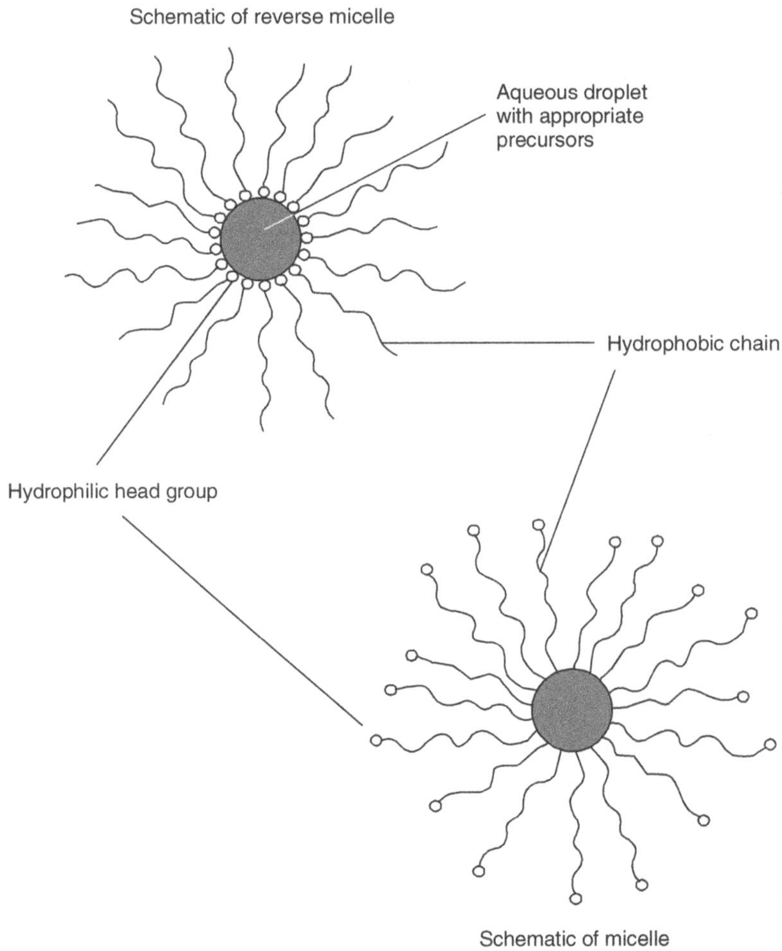

FIGURE 3.6 WATER-OIL MICROEMULSION (MICELLE) METHOD

A technique, which in some ways combines the aspects of different solution precipitation techniques, is the hydrodynamic cavitation process. Nanocrystalline oxide ceramic particles in the range 100 nm to a few microns have been produced by hydrodynamic processing in a microfluidizer. The method of producing oxide nanoparticles by the hydrodynamic cavitation process begins with the co-precipitation of the metal oxide components. The precipitated slurry stream is then drawn into a device where it is immediately elevated to high pressures within a small volume. The precipitated gel experiences ultrashear forces and cavitational heating. These two aspects lead to the formation of nanophase particles and high-phase purity in complex metal oxides.

COMMERCIAL PRODUCTION AND USE OF NANOPARTICLES

Over the past 5 years or so, nanoparticle producers have been working hard to differentiating themselves from their competitors, by either providing nanopowders with varied particle characteristics or by developing nanoparticles of proprietary compositions. In many instances, the ability to supply tonnage quantities of nanopowders has also been established. The selling price (which in many cases is a function of the manufacturing cost) has also come down over the years. Good quality alumina nanoparticles, that can be dispersed in a polymer matrix without substantially compromising the optical clarity, sell for ~\$100.00/kg. It should be noted that this is still an order of magnitude higher than what submicron size alumina sells for, which is ~\$10/kg. For obvious reasons, there has been no widespread replacement of micrograined materials by nanomaterials in applications. On the other hand, applications where a small amount, say approximately 10% wt, of nanoparticle addition has been able to change substantially the properties and performance of the end product, are becoming increasingly popular. A number of such examples can be found in the area of functional coatings.

CHARACTERIZATION OF NANOSTRUCTURE

LITHOGRAPHY TECHNIQUES

Originally, lithography was a printing method invented in 1798 by Alois Senefelder in Germany. At that time, there were only two printing techniques: relief printing and intaglio printing. In relief printing, a raised surface is inked and an image is taken from this surface by placing it in contact with paper or cloth. The intaglio process relies on marks engraved onto a plate to retain the ink. Lithography is based on the immiscibility of oil and water. Designs are drawn or painted with an oil-based substance (greasy ink or crayons) on specially prepared limestone. The stone is moistened with water, which the stone accepts in areas not covered by the crayon. An oily ink, applied with a roller, adheres only to the drawing and is repelled by the wet parts of the stone. The print is then made by pressing paper against the inked drawing. Optical lithography began in the early 1970s when Rick Dill developed a set of mathematical equations to describe the process of lithography. These equations published in the "Dill papers" marked the first time that lithography was described as a science and not an art. The first lithography modelling program SAMPLE was developed in 1979 by Andy Neureuther and Bill Oldman.

SPECIALIZED LITHOGRAPHY TECHNIQUES

In order to produce patterns at the nanometer scale, which is necessary for the fabrication of semiconductor integrated circuits, nanoelectromechanical systems, or lab-on-a-chip applications, specialized lithography techniques are used. Some of these techniques involve steps similar to those seen in photolithography; the differences lie in the use of energy sources with smaller wavelengths and smaller masks (both changes are used to produce nanoscaled patterns and structures). Such specialized lithography techniques include electron beam lithography, nanosphere lithography, and focused ion beam lithography (FIB). Other specialized techniques are more reminiscent of original lithography in that they transfer the pattern of molecules directly onto the substrate as a print. These techniques include types of soft lithography (such as microcontact printing, replica molding, microtransfer molding, and solvent-assisted micromolding), nanoimprint lithography, and dip-pen lithography (a type of scanning probe lithography).

Electron Beam Lithography

Electron beam lithography has been used in the production of semiconductors and the patterning of masks for other types of lithography (such as X-ray and optical lithography). In electron beam lithography, the exposed substrate is modified by the energy from a stream of electrons.

Nanosphere lithography Nanosphere lithography is similar to other types of lithography. In this type of lithography, the mask is replaced with a layer of nanospheres. After exposure and developing, the uncovered resin is washed away leaving behind nanoscale vertical columns.

Soft lithography Soft lithography is called "soft" because an elastomeric stamp or mould is the part that transfers patterns to the substrate and this method uses flexible organic molecules and materials rather than the rigid inorganic materials commonly used during the fabrication of microelectronic systems. Soft lithography can produce micropatterns of self-assembled monolayers (SAMs) through contact printing or form microstructures in materials through imprinting (embossing) or replica moulding. In this process, a self-assembled monolayer is stamped onto the substrate. The molecular impressions left by the monolayer can be used to seed crystal growth or bind strands of DNA for bioanalysis. Soft lithography techniques are not subject to the limitations set by optical diffraction (the edge definition is set by van der Waals interactions and the properties of the materials used). It is a procedurally simple, less expensive alternative to the production of nanoscaled structures through photolithography.

Dip-Pen Nanolithography

If we would control the architecture of a surface on the sub-100 nm and, preferably, down to the 1 nm length scale with reasonable speed and accuracy, some of the most important questions in science would be answered and, in the process, technologies that could allow for revolutionary advances in surface science and many related fields could be developed. Since the invention of scanning probe microscopes (SPMs) and the realization that one could manipulate matter in a one-atom-at-a-time fashion, many scientists thought that this elusive goal in controlling surface architecture might be possible. However, although early attempts to develop surface-patterning methodologies from SPMs were able to demonstrate the high resolution capabilities of these instruments, they were fraught with speed limitations or limitations with respect to the types of molecules and the conditions under which one could pattern. Indeed, throughout the 1980s and most of the 1990s, SPM surface patterning methods focused on either impressive, but inherently slow, serial scanning tunnelling microscope (STM) methods, that move individual atoms around on a surface under ultrahigh vacuum and low temperature conditions, or indirect multiple-step STM or atomic force microscope (AFM) etching ± backfilling methods. These approaches have the limitation that it is difficult to create parallel patterning methods based on them that are general with respect to controlling both feature size and the types of molecules that can be patterned within nanoscopic dimensions. In other words, one can utilize a multiple-tip SPM instrument with multiple independent feedback systems to control parallel etching procedures, but it is difficult, if not impossible, to selectively "fill in" such features with different types of molecules on the sub-100 nm length scale. In 1999, a new direct-write SPM-based lithographic method termed dip-pen nanolithography (DPN) was introduced to the scientific community. This method allows one to transport molecular substances to a surface, much like a macroscopic dip pen transfers ink to paper, but with the resolution of an AFM.

Two important observations made DPN possible. First, in studying the dynamics of the capillary effect and water transport from an AFM tip in air, it was realized that the meniscus could be used which naturally forms in air between the tip and sample, as a nanoscale reaction vessel and "ink" transport medium. This provides a way of controlling molecular transport between an ink-coated AFM tip and substrate. Second, if one used "inks" that were designed to react with the surface to be patterned, one would have a chemical driving force that would favour the movement of such inks from the AFM tip through the water-filled capillary to the substrate. Moreover, chemisorption of the ink onto the substrate would lead to stable, chemisorbed nanostructures. Initial DPN demonstrations showed that alkanethiols could be patterned onto gold surfaces and that the feature size and basic transport process could be controlled by regulating humidity and temperature. Hydrophilic and hydrophobic molecules, including biomolecules such as DNA, have now been patterned via the DPN process, and through the development of controlled environment chambers, which will allow one to replace the water meniscus with other solvents, it seems likely that it will be possible to extend DPN to a wide range of molecule types and substrates. One feature that sets DPN apart from other high resolution nanolithographic techniques is its nanostructure registration capabilities. Since with the DPN process, one uses the same high resolution tool to "read" and "write", one can generate multiple chemically pristine nanostructures, made of the same or different inks, and align them with respect to one another with near-perfect precision (less than 5 nm alignment resolution). This capability makes DPN an excellent customization tool for generating multifunctional nanostructures and for making chips with integrated nanostructures comprised of different chemical components, including biomolecules such as peptides and DNA. For this reason, DPN could have a major impact in molecular electronics, biodiagnostics, catalysis, and tribology.

With conventional cantilevers, DPN offers 15 nm linewidth and 5 nm spatial resolution and, therefore, rivals other techniques such as e-beam lithography for patterning solid substrates. The resolution of DPN will undoubtedly improve through the use of sharper cantilevers and as we improve our understanding of the "ink" transport process. Significantly, DPN has been transformed from a serial to parallel process through the use of eight cantilevers used as a single cantilever in a conventional AFM instrument. Consequently, a bundle of tips can be used in a parallel writing fashion simply by applying a contact force suitable to engage all of the tips in the bundle with the substrate to be patterned. Interestingly, each of these tips also can be engaged individually to allow one to use the multiple pen system in serial fashion like a nanoplotter; in the nanoplotter mode, one can use DPN to generate customized features comprised of multiple inks without the need to change tips during a patterning experiment. Rinsing and inking wells located on the periphery of the sample make DPN an almost totally automated process.

Although DPN is particularly well suited for patterning flat solid substrates with soft matter, when combined with wet chemical etching procedures, it also can be used to fabricate silicon structures, thus far on a length scale greater than 40 nm (Figure 3.7). Recently, silicon nanostructures were generated via a multistep process that utilizes DPN to generate soft etch-resistant structures comprised of octadecanethiol (ODT) on gold–titanium–silicon trilayer structures. The unmodified gold could be selectively etched with a ferri/ferrocycanide-based etching solution, to leave behind an alkanethiol-capped gold nanostructure. The exposed titanium-coated silicon subsequently could be anisotropically etched with KOH yielding raised trilayer Au–Ti–Si nanostructures. Finally, the Au and Ti layers can be removed by treatment with aqua region. This approach, in principle, could be extended to silver- or platinum-coated silicon substrates.

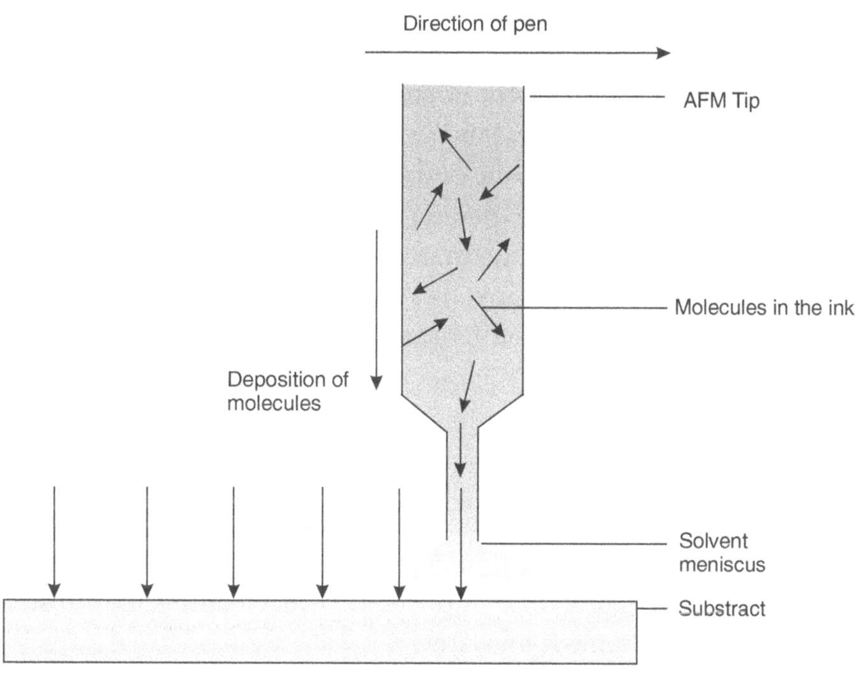

FIGURE 3.7 DIP-PEN LITHOGRAPHY, TRANSFERRING THE MOLECULES FROM THE PEN TO THE SUBSTRATE SURFACE

Although DPN is still in its infancy, its development already has led to the accomplishment of several important milestones in nanotechnology and in scanning probe lithographies in particular. First, it is now a user-friendly form of lithography accessible to any group with a conventional AFM. This alone should accelerate its use throughout the scientific community at large and especially within the chemical and biochemical communities. Second, its direct-write nature sets it apart from all scanning or physical processes. These include crystallization, catalysis, chemical and biochemical recognition, etching behaviour, surface molecular transport, and many others. One can determine the type of pattern appropriate for a given application and, because of the dimensions associated with these patterns, one will be able to use the same scanning probe instrument used to generate such structures to screen thousands of them on a relatively fast time scale (minutes). This combinatorial-nanotechnology approach could change the way we explore many nanoscale phenomena and, therefore, have a major impact on the development of new technologies based on materials with sub-100 nm dimensions.

PHOTOLITHOGRAPHY

Photolithography is a technique used to transfer shapes and designs on to the surface of photoresist materials. Over the years, this process has been refined and miniaturized; microlithography is currently used to produce items such as semiconductors for computers and an array of different biosensors. To date, photolithography has become one of the most successful technologies in the field of microfabrication. It has been used regularly in the semiconductor industry since the late 1950s; a great deal of integrated circuits has been manufactured by this technology. Photolithography involves several generalized steps, cleaning of the substrate, application of the photoresist material, soft baking, exposure, developing, and hard baking.

Cleaning of the Substrate

During substrate preparation, the material onto which the pattern will be developed is cleaned to remove anything that could interfere with the lithography process including particulate matter and impurities. After cleaning, the substrate is dried, usually in an oven, to remove all water.

Application of the Photoresist Material

There are two types of photoresist materials, positive and negative. Positive photoresists become more soluble when they are exposed to UV light. So in photolithography, when the mask is laid down onto the positive photoresist, the exposed areas (those not covered by the mask) will be removed by the developing solution leaving only the shape of the mask and the underlying substrate.

Negative photoresist materials work in the opposite manner. These photoresist materials polymerize on exposure to UV light, making them less soluble after exposure. Once the mask has been lifted and the material has been washed with developing solution, the photoresist covered by the mask is washed away. Therefore, using a negative photoresist creates the photographic negative of the mask. Photoresists commonly used in the production of microelectronics include ethylene glycol, monoethyl ether propylene glycol and methyl ether acetate.

Limitations of Photolithography

Current photolithography techniques used in microelectronics manufacturing use a projection printing system (known as a stepper). In this system, the image of the mask is reduced and projected via a high numerical aperture lens system, onto a thin film of photoresist that has been spin coated onto a wafer. The resolution that the stepper is capable of is based on optical diffraction limits set in the Rayleigh equation.

$$R = \kappa_1 \lambda / NA$$

In the Rayleigh equation, κ_1 is a constant that is dependent on the photoresist, λ is the wavelength of the light source, and NA is the numerical aperture of the lens. The minimum feature size that can be achieved with this technique is approximately the wavelength of the light used, λ, although theoretically, the lower limit is $\lambda/2$. So, in order to produce micro- or nanoscaled patterns and structures, light sources with shorter wavelengths must be used. This also makes manufacturing more difficult and expensive.

THIN FILM DEPOSITION

Thin film deposition techniques allow one to deposit a thin layer of material onto a surface or a substrate. Typically, the thicknesses of these thin films are in the nanometer scale. Some techniques even allow the addition of single atom layers of a material to the substrate. These techniques are used to create a wide variety of devices including coatings for optics components, conductors, semiconductors, and insulators for electronic devices, and films for packaging. Although thin film deposition includes a broad range of technologies and methodologies, a generalized process can be described in three steps: 1) creation of the flux of condensable species–this includes either neutral atoms or ions, 2) transportation of the created species to the substrate, and

3) growth of the film onto the substrate. Thin film deposition has a wide variety of applications including the manufacture of reflective coatings for optics, the manufacture of electronics, and the purification of metals such as copper through electroplating. Electronics is perhaps the most well-known field that requires thin film deposition. The technology is continually allowing the electronics industry to create smaller and smaller electronic components including insulators, semiconductors, and integrated circuits, allowing developers to increase the speed and efficiency of their products without increasing the product size. Thin film deposition can be divided into two broad categories: physical deposition and chemical deposition. Only a few thin film deposition techniques do not fit completely into these categories; they will be described later. Many methods of thin film deposition do not rely on one technology to attain their goals; often a combination of techniques and technologies are used.

Physical deposition, as the name implies, involves thin film deposition techniques that use physics, typically in terms of mechanical or thermodynamic means, to deposit the desired film onto the substrate. Physical deposition includes three primary classes: evaporation, sputtering, and ion beam. Evaporation thin film deposition employs the ability to evaporate the material to be deposited by physical heating to create the flux of material. The species created through the heating process is then transported to the substrate to be deposited onto the surface. Although rather simple in concept, many different methods to heat the materials have been used, including resistance heating, electron beam heating, arc evaporation, induction heating, and flash evaporation. An example of arc evaporation is cathodic arc plasma deposition (CAPD). This technique utilizes vacuum arcs to produce the condensable species. The source material to be deposited is created from the cathode in the arc discharge circuit. The condensable species is created from the source material through flash evaporation as the arc spots move along the surface of the target. The arc spots are sustained by the plasma being formed by the arc itself.

The second class of physical thin film deposition is the sputtering technique. This process involves the synthesis of the condensation species by bombardment of the source (target) with positive ions of an inert gas. The collisions cause the atoms on the target to be knocked off through momentum transfer and result in the creation of ions on the target as well. This technique is typically called plasma-assisted (enhanced) deposition, which encourages the atoms to be knocked off and collected onto substrates. The sputtering technique has an advantage over the aforementioned evaporation technique because it can occur at lower temperatures (heating is not required) and because the process is independent of the evaporation rate of the material.

Another class of physical thin film deposition utilizes ion beam technology. These techniques employ the use of an ion beam to ionize the material to be deposited. An example is gas cluster ion beam technology (GCIB). Neutral gas clusters are produced by the expansion of atoms or molecules at high pressure through a room temperature nozzle into a vacuum. The neutral gas cluster is ionized by the bombardment of electrons and is accelerated by a high voltage to impact a substrate. The impact causes all of the atoms to nearly interact simultaneously and deposit a very high energy density into a small volume of the target material.

The second large category of thin film deposition techniques is chemical deposition. As the name implies, chemical deposition causes a chemical change of the fluid material that results in the deposition. Chemical vapour deposition (CVD) is a class of chemical deposition where the condensable species is formed from gases or vapours that are, without energy input, not condensable. The substrates are heated at high temperature to cause the gases to decompose resulting in deposition. Plasma-enhanced chemical vapour deposition (PECVD) utilizes plasma that is created through the action of an electric field. Ionization, dissociation, and gas phase reactions occur as the reactant gases are passed through the low pressure plasma. The addition of plasma to the technique allows the process to occur at much lower temperatures, not entirely relying on the reactions to be driven thermally.

Plating is another class of chemical deposition used to produce thin films. The plating technique distinguishes itself because of its use of liquid precursors. The material to be deposited is initially dissolved in an aqueous solution. Although the process can be driven entirely by the reagents within the solution, usually an electric current or ion beam is used to drive the reaction to form the layer of desired material.

The final two techniques of thin film deposition to be discussed are molecular beam epitaxy (MBE) and reactive sputtering. MBE and reactive sputtering do not entirely fit into either chemical or physical deposition categories, being more of a combination of chemical and physical. During MBE, ultrapure elements are heated until they begin to slowly evaporate. The evaporated elements do not react with one another until they condense onto the substrate. The process occurs in a high vacuum, which allows a slow deposition rate. The slow deposition rate allows the material to grow epitaxially, meaning that instead of many randomly arranged grains being formed, as in higher deposition rates, the material tends to grow in larger grains with a more uniform orientation. The slow deposition rate also allows the element to be deposited one layer at a time.

The last technique to be discussed is reactive sputtering. In reactive sputtering, a small amount of non-noble gas (such as oxygen or nitrogen) is mixed with the plasma-forming gas. The material is then sputtered from the target to the substrate reacting with the gas. The result is the deposition of a different material. If the gas used is either oxygen or nitrogen, an oxide or nitride can be formed. The method is not limited to oxygen or nitrogen; a wide variety of compounds can be created via this method.

ELECTROSPINNING

Electrospinning is a technique used to create polymeric fibres with diameters in the nanometer range. This process involves the ejection of a charged polymer fluid onto an oppositely charged surface. One of the first investigations into the flow of conducting liquid through a charged tube with a counter electrode some distance away was conducted by Zeleny. A great deal of these experiments used aqueous electrolyte solutions with high electrical conductivity and low viscosity. The addition of a charge to this solution caused it to form a fine spray of charged droplets that were attracted to the counterelectrode. These droplets quickly evaporated in the air; this process was later called electrospraying. Electrospinning is very similar to electrospraying; a charge is applied to a polymer solution or melt and ejected toward an oppositely charged target. A typical electrospinning experimental set-up is shown in Figure 3.8. In both processes, the surface tension and viscoelastic forces of the polymer solution cause drops at the tip of the syringe to retain their hemispherical shape. The charge induced by the electric field causes the droplet to deform into a "Taylor cone" at the tip of the tube.

When the applied voltage is increased beyond a threshold value, the electric forces in the droplet overcome the opposing surface tension forces and a narrow charged jet is ejected from the tip of the Taylor cone. In electrospraying, the strength of the electric field and low viscosity of a solution cause droplets to separate from

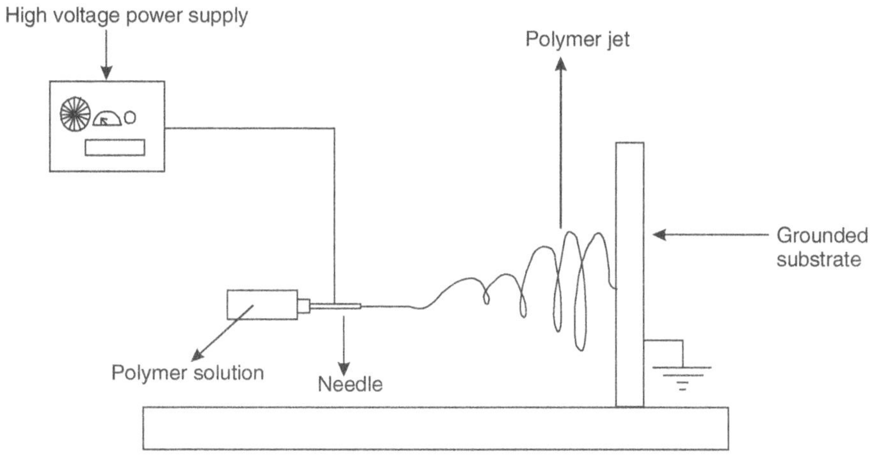

FIGURE 3.8 A TYPICAL SET-UP OF ELECTROSPINNING

the cone and spray onto the target. In electrospinning, the concentration of the solution, viscosity of the solution, and the entanglement of the polymer chains cause a fibre to be extruded from the tip of the cone. The polymer jet begins as a nearly straight line because of the stabilization from the longitudinal stress of the external electrical field on the charge carried by the jet. However, the polymer jet quickly becomes unstable because of the repulsive forces from the opposite charges in the polymer jet. The jet experiences electrically driven bending instabilities and the jet is whipped around in spiral. The polymer jet is stretched while it bends and travels through the spiral causing a significant decrease in fibre diameter and rapid evaporation of the solvent resulting in the formation of nanoscale thin fibres. These fibres land randomly on to the grounded target forming a nonwoven fibre mat.

Electrospinning produces fibres with submicron diameters. The diameter of the fibres can be adjusted by altering the distance between the polymer source and the target, the polymer concentration, and the voltage. The orientation of the fibres onto the target can be changed by employing various external mechanical or electrostatic forces. The mechanical properties of the fibre mat can be altered by changing the diameter and orientation of the fibres.

Cellulose acetate was one of the first solutions electrospun; in another early study, molten wax was observed to form threads when subjected to high electrical field. Presently, a wide variety of polymers have been electrospun from natural polymers such as type I collagen or elastin to synthetic polymers such as polyphosphazenes. Several parameters affect the electrospinning process.

These parameters can be split into different groups, polymer solution properties, processing parameters, and environmental parameters. Polymer solution properties include solution viscosity, polymer molecular weight, polymer concentration, surface tension, solution conductivity, and dielectric effect of the solvent. Processing parameters that affect electrospinning include voltage, distance from needle to collector, flow rate, needle/orifice diameter, and type of collector. The environmental parameters that can affect electrospinning of nanofibres include humidity, pressure, and type of atmosphere.

NANOSPHERES

Nanosphere technology revolves around the creation of polymeric spheres that are hundreds of nanometres in diameter. Their tiny size and ability to absorb or bind to (through surface chemistry) drugs or growth factors make them ideal vehicles for drug or growth factor delivery. Polymeric nanospheres can be created by emulsion solvent diffusion. The polymer is dissolved into a volatile organic solvent (such as chloroform, ethyl acetate, or methylene chloride). This solution is then poured into an aqueous phase with a surfactant or stabilizer and agitated usually with a sonicator. The nanospheres are then collected by evaporating the solvent (this may be done with mixing) or dilution with water. Drugs and growth factors may be added to the organic or aqueous phase in order to incorporate them into the nanospheres. Scanning electron microscopy is used for nanosphere characterization. The size and density of the spheres can be altered by changing the amount of polymer added to the surfactant or stabilizer solution and the speed with which the solution is agitated.

CARBON NANOTUBES

Carbon is the building block for all organic life. It is found in a variety of structures that display vastly different properties. In nature, pure carbon is found in a variety of forms from the soft, flexible, and conductive graphite to the hard, inflexible insulating diamond. The properties of these materials stem from the arrangement of their carbon atoms. In graphite, the carbon atoms are arranged into sheets of honeycomb-shaped matrices; these matrices are held together by covalent bonds between the carbon atoms. The sheets are held to one another by weak van der Waals forces; this allows graphite to flake off into layers (sheets of carbon atoms slide over one another when force is applied) making graphite a good lubricant. In contrast, the carbon atoms in a diamond are arranged into a tetrahedral pattern.

This arrangement makes diamonds hard and prevents them from flaking like graphite. Similarly, it is the arrangement of their atoms and their nanoscale size that give carbon nanotubes their mechanical, chemical, and electrical properties.

Carbon nanotubes (CNTs) are fullerenes that are structurally similar to long rolls of graphite. Fullerenes are molecules composed entirely of carbon that exist as hollow objects (spheres, ellipsoids, or tubes); they are named after Buckminster Fuller, an American architect, designer, and author known for his work on geodesic domes. CNTs are created as single-walled nanotubes (SWNTs) or multiwalled nanotubes (MWNTs). SWNTs have only one wall whereas MWNTs have many walls formed around one another. The walls of MWNTs are held to one another by van der Waals forces. The properties of CNTs depend on their length, diameter, and chirality (the direction in which the graphite sheet is rolled up).

Carbon filaments were first noticed under gaseous conditions in the presence of metal catalysts in the 1950s after the development of the electron microscope. Carbon filaments up to several hundred nanometres long were noticed extending from catalytic particles of transition metals. Tibbetts was the first to document these structures and speculate about their growth mechanism. In the 1990s, researchers observed the formation of MWNTs through the electric arc discharge method and later documented the synthesis of SWNTs via laser ablation. SWNTs are very strong and very stiff because of the strength of the carbon–carbon bond and the seamless structure of the nanotube; the modulus of an SWNT has been calculated as 0.64 TPa with an ultimate tensile stress of approximately 37 GPa. The electrical properties of SWNTs can range from metallic to semiconducting with various band gaps depending on the chirality and diameter of the tube. CNTs have been used in a variety of applications including increasing mechanical properties of polymers through composite reinforcement, use in nanosensing devices, and in bioimaging. CNTs have been made using a variety of fabrication methods. These methods include electric arc discharge, laser ablation of carbon, and chemical vapour deposition, techniques that are also used to create thin films.

SELF-ASSEMBLED NANOSTRUCTURES

The self-assembly of molecules into structures is a phenomena seen frequently in nature; type I collagen molecules are examples of molecules that can orient and arrange themselves to form two-dimensional and three-dimensional structures. Among the nanofabrication methods discussed in this chapter, self-assembly of molecules is one of the most promising ways to form a large variety of nanostructures. This technique relies on the noncovalent interactions between molecules (electrostatic, van der Waals, hydrogen bonding, π–π interactions, and capillary force) to organize groups of molecules into larger, regular structures. Self-assembled monolayers are among the most commonly studied self-assembled structures. SAMs are surfaces covered by a thin film consisting of a single layer of molecules. Typically, they are formed when surfactant molecules are absorbed onto a substrate as a monomolecular layer. Among the most common and widely studied SAM systems are gold-alkylthiolate ($CH_3(CH_2)_nS$) and silicon oxide-alkylsilane monolayer systems (Figure 3.9).

In typical SAM fabrication, a group with a strong attraction to a particular substrate is attached to the head of a long molecule, an alkane chain with about 10–20 methylene units. The head group absorbs easily onto the substrate surface, creating a monolayer with the tails pointing away from the substrate surface. An example of this is the gold-thiol SAM; thiol (S — H) head groups in solution absorb readily onto gold. Next, a layer of gold is added in order to allow the thiol groups to bind. The wafers are covered in gold using vapour deposition of chromium and gold from heated tungsten boats in a cryogenically pumped deposition chamber.

The layer of chromium is deposited before the layer of gold to enhance adhesion to the substrate. The gold-coated samples are removed and placed into a 1 mM solution of 1-hexadecanethiol in ethanol, contained at ambient temperature, for 3 days. Afterward, the wafers are placed into the thiol solution. Before experimental use, they are rinsed with ethanol and air-dried. In this method, the thiol heads form a dense layer (1 molecule thick) on the gold surface whereas the tails of the molecules point outward forming another layer covering the substrate surface.

By changing the group attached to the tail of the absorbed molecules (or using different molecules), we can tailor the functionality of the new surface. Functionality can also be added to that tail groups after the SAM is formed. The tail groups and molecules can be selected to form SAMs for specific applications. Adding short oligomers of ethylene glycol groups to the tail decreases protein adhesion. Methyl-terminated groups can increase plasma and protein binding, which could increase the biocompatibility of the substrate. Absorbing a layer of octadecyltrichlorosilane (OTS) increases wear resistance of polysilicon, a technique that may increase the durability of MEMS or NEMS.

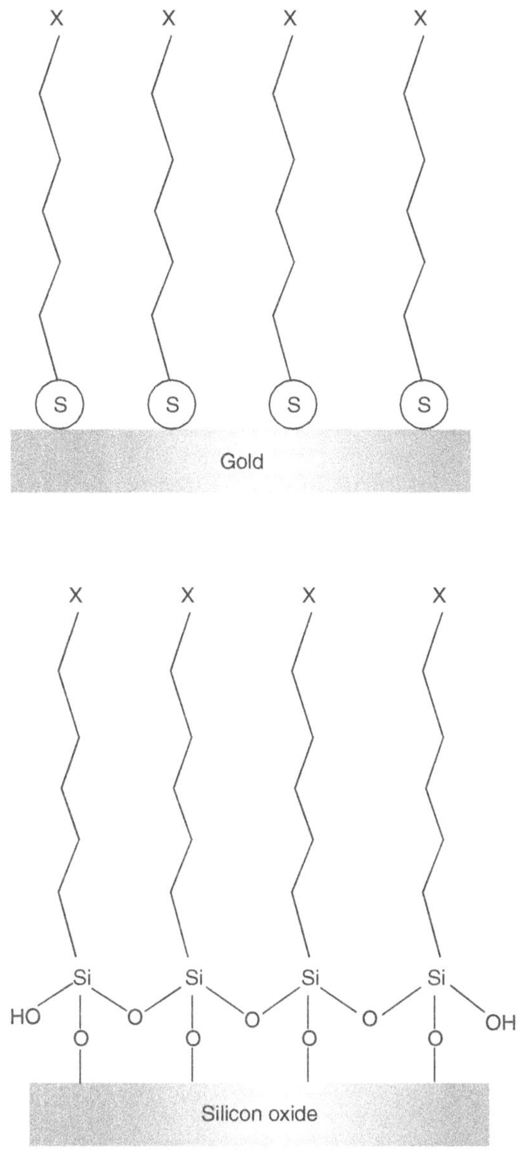

FIGURE 3.9 A GOLD-ALKYLTHIOLATE AND SILICON OXIDE-ALKYLSILANE

ATOMIC FORCE MICROSCOPE

Within the field of scanning probe microscopy atomic force microscopy (AFM) is extensively used in a wide range of disciplines such as life science, solid-state physics, and materials science. The AFM has evolved into an imaging method that yields structural details of biological samples such as proteins, nucleic acids, membranes, and cells in their native environment. AFM is a unique technique for providing subnanometer resolution at a reasonable signal-to-noise ratio under physiological conditions. As a result of continuous developments in sample preparation, imaging techniques, and instrumentation, AFM is now a companion technique to X-ray crystallography and electron microscopy (EM) for the determination of protein structures, for example. It complements EM by allowing visualization of biological samples in buffers that preserve their native structure over extended time periods. AFM does not rely on symmetry averaging and crystallization, therefore revealing defects and structural anomalies not observable in classical ensemble measurements. Unlike EM, AFM yields three-dimensional maps with an exceptionally good vertical resolution (less than a nanometer).

In addition to high-resolution imaging of proteins, nucleotides, membranes, and living cells, the measurement of mechanical forces at the molecular level provides detailed insights into the function and structure of biomolecular systems. Inter- and intramolecular interactions can be studied directly at the molecular level, as exemplified by the analysis of polysaccharide elasticity, DNA mechanics, the function of molecular motors, and the binding potentials of receptor–ligand pairs involved in cell adhesion. In the latter case, defined forces are exerted on a receptor–ligand complex and the dissociation process is followed over time. With data obtained from force measurements, molecular interactions can be analysed in terms of kinetic rate constants, structural parameters of binding pockets, molecular dynamics of the recognition process, and the energy landscape of the interaction. Compared with conventional ensemble methods, single-molecule experiments offer several advantages.

First, by conducting many sequential measurements, the distribution of molecular properties of inhomogeneous systems can be determined. Second, single-molecule trajectories, since they are direct records of the system's fluctuations, provide dynamic and statistical information that is often hidden in ensemble-averaged results. Finally, it is possible to monitor rarely populated transients in real time, which are difficult or impossible to capture using conventional methods.

High-Resolution Imaging of Biological and Nanostructured Materials

Understanding how complexes of nanostructures are assembled under certain conditions is of fundamental importance to elucidate their function. Accordingly, considerable attention has been paid to high-resolution topographical imaging of individual molecules and their complexes. The AFM operates by scanning a pointed probe over a surface while adjusting the force applied to the sample to minimal values that prevent displacement or destruction of weakly adhering nanostructured materials. Problems arising from unfavourable probe–surface interactions in contact mode AFM, especially lateral forces, have been largely overcome by the development of dynamic force microscopy (DFM) methods, mainly employed for the investigation of soft biological membranes and supramolecular complexes. In DFM, the cantilever is oscillated close to its resonance frequency at an amplitude of a few nanometres as it raster-scans the surface and touches the sample only at the end of its downward movement. By taking advantage of direct cantilever excitation in magnetic AC mode AFM (where tip oscillation is accomplished by an external magnetic field), lateral forces can be minimized and the tip oscillation properly controlled. Researchers have recently demonstrated that DFM even allows the imaging of hard protein surfaces with similar resolution to contact mode AFM.

A prerequisite for obtaining high-quality AFM images of biological samples is tight immobilization on a flat support. The simplest and most common immobilization technique is physical adsorption from solution. Muscovite mica, the most frequently used support for biological applications, is a layered mineral that can be easily cleaved, leading to clean, atomically flat surfaces. Because the mica surface is negatively charged at neutral pH, simple adsorption of positively charged samples is sufficient for stable immobilization. Even negatively charged molecules (e.g. DNA) can be physically adsorbed onto mica surfaces when multivalent cations (like Ni^{2+}, Mg^{2+}, Zn^{2+}) are present in buffer solution. DFM has also been used to study the surface of human rhinovirus (HRV) under physiological conditions. HRVs belong to one of the largest and most important families of viral pathogens, the picornaviruses, which cause the common cold. For specific and tight immobilization of virus particles, a method of assembling HRV on artificial membrane interfaces that mimic the cell surface is used. A supported lipid bilayer containing Ni^{2+}–nitrilotriacetate (NTA) lipids as functional groups in the top layer at high surface density is assembled onto mica. The specific interaction between Ni^{2+}–NTA lipids and soluble His_6-tagged virus receptors is used to assemble a planar matrix of laterally diffusing receptor proteins. After injection of HRV, a two-dimensional crystal of hexagonally arranged virus particles forms on the receptor surface. Topographical imaging of the virus capsid reveals a regular arrangement of 3-nm-sized protrusions similar to that seen in cryo-electron microscopy studies.

For studying biomolecular assemblies, DFM can be used to follow antibody binding to surface antigens embedded in cell membranes, which is the primary event in the specific immune defence of vertebrates.

In a series of AFM images, the *in situ* binding of individual antibodies can be followed and visualized over time, with kinetic binding parameters inferred at the single molecule level. Fab fragments (i.e., the fragment antigen-binding region in antibodies) of membrane-bound antibodies are allocated to antigenic sites at 1.5 nm lateral resolution on purple membrane, allowing the identification and localization of individual recognition sites (epitopes).

FOURIER TRANSFORM INFRARED SPECTROSCOPY (FT-IR)

FT-IR stands for Fourier Transform InfraRed, the preferred method of infrared spectroscopy. In infrared spectroscopy, IR radiation is passed through a sample. Some of the infrared radiation is absorbed by the sample and some of it is passed through (transmitted). The resulting spectrum represents the molecular absorption and transmission, creating a molecular fingerprint of the sample. Like a fingerprint no two unique molecular structures produce the same infrared spectrum. This makes infrared spectroscopy useful for several types of analysis. The following information that we can get from FT-IR.

- It can identify unknown materials
- It can determine the quality or consistency of a sample
- It can determine the amount of components in a mixture

Importance of Infrared Spectroscopy

Infrared spectroscopy has been a workhorse technique for materials analysis in the laboratory for over seventy years. An infrared spectrum represents a fingerprint of a sample with absorption peaks which correspond to the frequencies of vibrations between the bonds of the atoms making up the material. Because each different material is a unique combination of atoms, no two compounds produce the exact same infrared spectrum. Therefore, infrared spectroscopy can result in a positive identification (qualitative analysis) of every different kind of material. In addition, the size of the peaks in the spectrum is a direct indication of the amount of material present. With modern software algorithms, infrared is an excellent tool for quantitative analysis.

Older Technology

The original infrared instruments were of the dispersive type. These instruments separated the individual frequencies of energy emitted from the infrared source. This was accomplished by the use of a prism or grating. An infrared prism works exactly the same as a visible prism which separates visible light into its colours (frequencies). A grating is a more modern dispersive element which better separates the frequencies of infrared energy. The detector measures the amount of energy at each frequency which has passed through the sample. This results in a spectrum which is a plot of intensity vs. frequency. Fourier transform infrared spectroscopy is preferred over dispersive or filter methods of infrared spectral analysis for several reasons.

- It is a non-destructive technique.
- It provides a precise measurement method which requires no external calibration.
- It can increase speed, collecting a scan every second.
- It can increase sensitivity—one-second scans can be co-added together to ratio out random noise.
- It has greater optical throughput.
- It is mechanically simple with only one moving part.

NEED FOR FT-IR

Fourier Transform Infrared (FT-IR) spectrometry was developed in order to overcome the limitations encountered with dispersive instruments. The main difficulty was the slow scanning process. A method for measuring all of the infrared frequencies simultaneously, rather than individually, was needed. A solution was developed which employed a very simple optical device called an interferometer. The interferometer produces a unique type of signal which has all of the infrared frequencies "encoded" into it. The signal can be measured very quickly, usually on the order of one second or so. Thus, the time element per sample is reduced to a matter of a few seconds rather than several minutes.

Most interferometers employ a beamsplitter which takes the incoming infrared beam and divides it into two optical beams. One beam reflects off of a flat mirror which is fixed in place. The other beam reflects off of a flat mirror which is on a mechanism which allows this mirror to move a very short distance (typically a few millimetres) away from the beamsplitter. The two beams reflect off of their respective mirrors and are recombined when they meet back at the beamsplitter. Because the path that one beam travels is a fixed length and the other is constantly changing as its mirror moves, the signal which exits the interferometer is the result of these two beams "interfering" with each other. The resulting signal is called an interferogram which has the unique property that every data point (a function of the moving mirror position) which makes up the signal has information about every infrared frequency which comes from the source. This means that as the interferogram is measured, all frequencies are being measured simultaneously. Thus, the use of the interferometer results in extremely fast measurements. Because the analyst requires a frequency spectrum (a plot of the intensity at each individual frequency) in order to make an identification, the measured interferogram signal can not be interpreted directly. A means of "decoding" the individual frequencies is required. This can be accomplished via a well-known mathematical technique called the Fourier transformation. This transformation is performed by the computer which then presents the user with the desired spectral information for analysis.

The Sample Analysis Process

The normal instrumental process is as follows:

The source Infrared energy is emitted from a glowing black-body source. This beam passes through an aperture which controls the amount of energy presented to the sample (and, ultimately, to the detector).

The interferometer The beam enters the interferometer where the "spectral encoding" takes place. The resulting interferogram signal then exits the interferometer.

The sample The beam enters the sample compartment where it is transmitted through or reflected off of the surface of the sample, depending on the type of analysis being accomplished. This is where specific frequencies of energy, which are uniquely characteristic of the sample, are absorbed.

The detector The beam finally passes to the detector for final measurement. The detectors used are specially designed to measure the special interferogram signal.

The computer The measured signal is digitized and sent to the computer where the Fourier transformation takes place. The final infrared spectrum is then presented to the user for interpretation and any further manipulation.

Because there needs to be a relative scale for the absorption intensity, a background spectrum must also be measured. This is normally a measurement with no sample in the beam. This can be compared to the measurement with the sample in the beam to determine the "per cent transmittance." This technique results in a spectrum which has all of the instrumental characteristics removed. Thus, all spectral features which are present are strictly due to the sample. A single background measurement can be used for many sample measurements because this spectrum is characteristic of the instrument itself.

Advantages of FT-IR

Some of the major advantages of FT-IR over the dispersive technique include:

Speed Because all of the frequencies are measured simultaneously, most measurements by FT-IR are made in a matter of seconds rather than several minutes. This is sometimes referred to as the Felgett Advantage.

Sensitivity Sensitivity is dramatically improved with FT-IR for many reasons. The detectors employed are much more sensitive, the optical throughput is much higher (referred to as the Jacquinot Advantage) which results in much lower noise levels, and the fast scans enable the co-addition of several scans in order to reduce the random measurement noise to any desired level (referred to as signal averaging).

Mechanical simplicity The moving mirror in the interferometer is the only continuously moving part in the instrument. Thus, there is very little possibility of mechanical breakdown.

Internally calibrated These instruments employ a He-Ne laser as an internal wavelength calibration standard (referred to as the Connes Advantage). These instruments are self-calibrating and never need to be calibrated by the user.

These advantages, along with several others, make measurements made by FT-IR extremely accurate and reproducible. Thus, it a very reliable technique for positive identification of virtually any sample. The sensitivity benefits enable identification of even the smallest of contaminants. This makes FT-IR an invaluable tool for quality control or quality assurance applications whether it be batch-to-batch comparisons to quality standards or analysis of an unknown contaminant. In addition, the sensitivity and accuracy of FT-IR detectors, along with a wide variety of software algorithms, have dramatically increased the practical use of infrared for quantitative analysis.

Quantitative methods can be easily developed and calibrated and can be incorporated into simple procedures for routine analysis. Thus, the Fourier Transform Infrared (FT-IR) technique has brought significant practical advantages to infrared spectroscopy. It has made possible the development of many new sampling techniques which were designed to tackle challenging problems which were impossible by older technology. It has made the use of infrared analysis virtually limitless.

DIFFERENTIAL SCANNING CALORIMETRY (DSC)

Differential Scanning Calorimetry or DSC is a thermoanalytical technique in which the difference in the amount of heat required to increase the temperature of a sample and reference are measured as a function of temperature. Both the sample and reference are maintained at nearly the same temperature throughout the experiment. Generally, the temperature program for a DSC analysis is designed such that the sample holder temperature increases linearly as a function of time. The reference sample should have a well-defined heat capacity over the range of temperatures to be scanned.

Differential scanning calorimetry can be used to measure a number of characteristic properties of a sample. Using this technique it is possible to observe fusion and crystallization events as well as glass transition temperatures (T_g). DSC can also be used to study oxidation, as well as other chemical reactions. Glass transitions may occur as the temperature of an amorphous solid is increased. These transitions appear as a step in the baseline of the recorded DSC signal. This is due to the sample undergoing a change in heat capacity; no formal phase change occurs.

As the temperature increases, an amorphous solid will become less viscous. At some point the molecules may obtain enough freedom of motion to spontaneously arrange themselves into a crystalline form. This is known as the crystallization temperature (T_c). This transition from amorphous solid to crystalline solid is an exothermic process, and results in a peak in the DSC signal. As the temperature increases the sample eventually reaches its melting temperature (T_m). The melting process results in an endothermic peak in the DSC curve. The ability to determine transition temperatures and enthalpies makes DSC an invaluable tool in producing phase diagrams for various chemical systems. Figure 3.10 shows a schematic of DSC curve demonstrating the appearance of several common features.

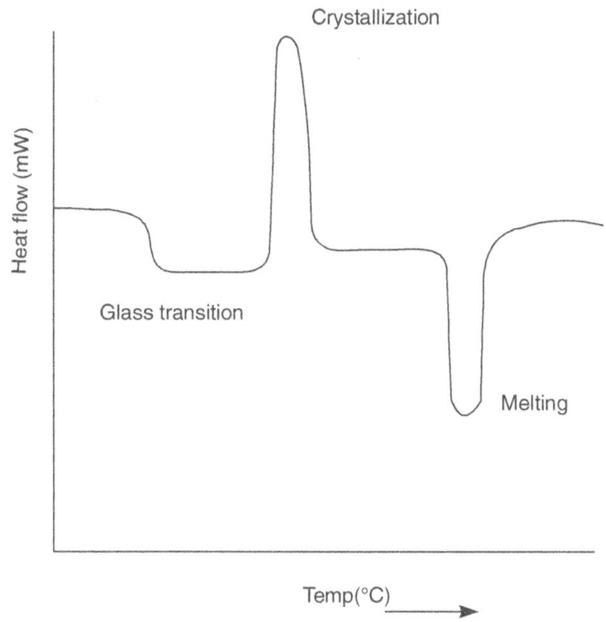

FIGURE 3.10 FEATURE OF DIFFERENTIAL SCANNING CALORIMETRY CURVE

POLYMERS

DSC curves may also be used to evaluate drug and polymer purities. This is possible because the temperature range over which a mixture of compounds melts is dependent on their relative amounts. This effect is due to a phenomenon known as freezing point depression, which occurs when a foreign solute is added to a solution. (Freezing point depression is what allows salt to de-ice sidewalks and antifreeze to keep your car running in the winter.) Consequently, less pure compounds will exhibit a broadened melting peak that begins at lower temperature than a pure compound.

DSC is used widely for examining polymers to check their purity and composition. Melting points and glass transition temperatures for most polymers are available from standard compilations, and the method can show up possible polymer degradation by the lowering of the expected melting point, T_m, for example. T_m depends on the molecular weight of the polymer, so lower grades will have lower melting points than expected. In the pharmaceutical industry it is necessary to have well-characterized drug compounds in order to define processing parameters. For instance, if it is necessary to deliver a drug in the amorphous form, it is desirable to process the drug at temperatures below those at which crystallization can occur.

CLOSING REMARKS

Nanotechnology is an exciting field that has seen tremendous growth and has advanced into the biomedical and electronics arena. Nanotechnology has become an integrated part of our everyday lives. From medicine, to clothing, to electronics, breakthroughs in nanofabrication have lead to new products. There are a plethora of techniques for creating nanoscaled devices for biomedical engineering. Many of these, such as lithography-based techniques, are applications of older techniques with more modern and powerful equipment. All of these techniques produce patterns, structures, and surface morphologies at the nanoscale that lead to devices and materials with unique surface properties and functionality. These techniques impart unique chemical, mechanical, electrical, and optical properties to materials for use in electronics, materials science, and medicine. Nanofabrication has lead to major advances in these fields and holds promise for the creation of new technologies.

CONCEPT CHECK

1. Is DSC study for qualitative or quantitative analysis?
2. How will you analyse the unknown material with FT-IR?
3. Mention the name of any two commercially available liposome or micelle.
4. Which is the most frequently used method for nanoparticle synthesis?
5. What to do you mean by sol-gel process?
6. What is the difference between micelle and liposomes?
7. What is the role of water-in-oil composition in micelle?
8. Define FJB.
9. Give the uses of DPN.
10. What are the limitations behind the photolithography process?
11. Differentiate CVD from PECVD.
12. What is the contribution of Zeleny?
13. What do you mean by "Taylor cone"?

14. What is the role of van der Waals force in MWNT?
15. What is the role of muscovite mica?
16. Define interferometer and interferogram.
17. What is "Felgett advantage"?
18. Define grain boundaries.

MIND TWISTER

1. How will you predict the efficiency of drug encapsulation from polymer with the help of DSC?

REFERENCES

Henshaw, M. (2003). First impressions: the early history of lithography: a comparative survey. *Artonview*. 33:33–38.

Hong, S. and Mirkin, C.A. (2000). Dip-pen Nanolithography: direct writing soft structures on the sub-100 nanometer-length scale. *Science*. 288, 1808.

Hong, S. Zhu, J. and Mirkin, C.A. (1999). Size distribution of particulate sulfate, nitrate and ammonium at a coastal site in Hong Kong. *Science*. 286, 523.

Huang, Z.M., Zhang, Y.Z., Koraki, M. and Ramakrishna, S. (2003). A review on polymer nanofibers by electrospinning and their applications in nanocomposites. *Composites Sci. Technol.* 63:2223–2253.

Nikova, D. *et al.* (2006). *Applied Scanning Probe Methods*. Vol. III. Springer, Berlin. pp.117–132.

Weinberger, D.A., Hong, S.H., Mirkin, C.A., Wessels, B.W. and Higgins, T. B. (2000). Combinatorial generation and analysis of nanometer- and micrometer-scale silicon features via dip-pen nanolithography and wet chemical etching. *Adv. Mater.* 12, 1600.

Xia, Y.N. and Whitesides, G.M. (1998). Soft lithography. *Annu. Rev. Mater. Sci.* 28: 158–184.

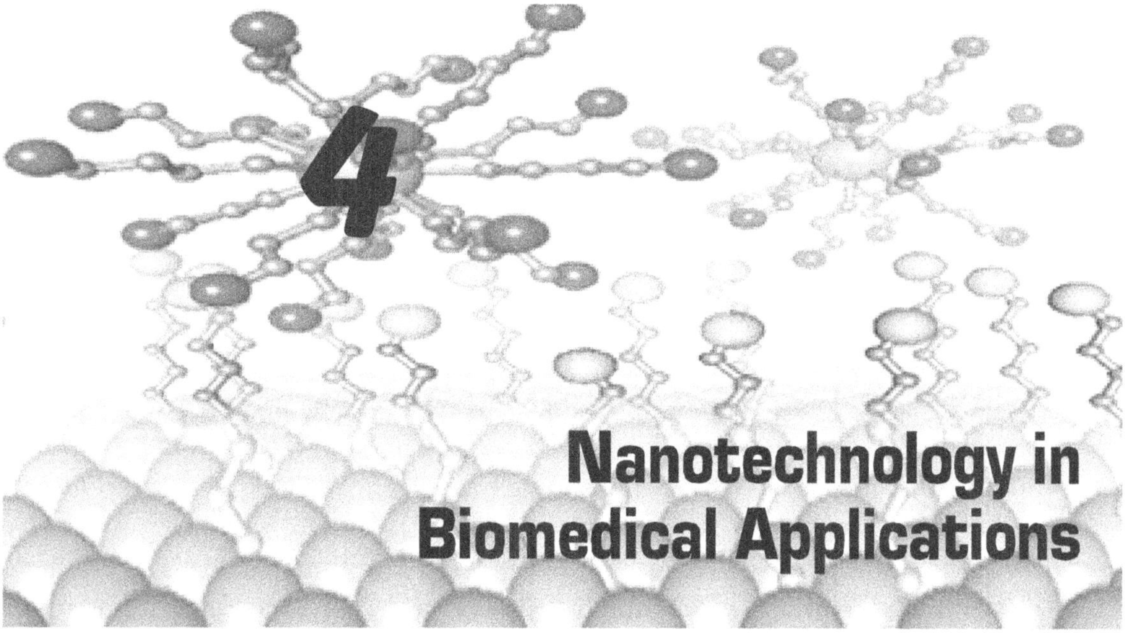

4 Nanotechnology in Biomedical Applications

INTRODUCTION

In the past 30 years, the explosive growth of nanotechnology has burst into challenging innovations in pharmacology, which is in the process of revolutionizing the delivery of biologically active compounds. The main input of today's nanotechnology in pharmacology is that it allows real progresses to achieve temporal and spatial site-specific delivery. Thus, the concept of the "magic bullet" proposed a century ago by the immunologist Nobel laureate Paul Ehrlich turned out recently to reality with the appearance of several approved forms of drug-targeting systems for the treatment of certain cancer and serious infectious diseases. This breakthrough was made possible by the development of various types of nanosystems resulting from cutting-edge researches based on pluridisciplinary approaches.

Nanosystems were also found useful to improve the performance of imaging techniques applied for the *in vivo* diagnosis of tumours. In this case, colloid metals are often incorporated in the nanodevice. Applications of nanotechnology in pharmacology are now undeniably linked to the potential of drug targeting. Although we are still far from the ideal "magic bullet", today, nanotechnology has already completed several key achievements to reach this goal. Another very demanding field includes infectious diseases [human immunodeficiency virus (HIV), leishmaniasis, malaria, nosocomial infections, all kinds of infections in immunocompromised patients, etc.] with already approved drugs for clinical uses (Ambisome®). Treatments of these severe diseases generally involved highly toxic compounds for healthy tissues, and their uses in therapy are considerably limited by occurrence of dramatic side effects using the traditional pharmaceutical formulations.

Nanotechnology also seems to be a promising alternative to overcome the problems of the administration of peptides and proteins and of the new drug molecules that are being discovered. Many of them are, indeed, poorly soluble in both aqueous and organic media, which results in poor bioavailability with low and/or erratic absorption when using traditional formulations. Nucleic acids are other potential candidates for which nanotechnology represents a unique opportunity to be used in therapy. They are rapidly degraded in biological media, and they hardly cross biological barriers. In addition, these molecules, which are even sometimes considered as 'undeliverable' compounds, need to reach an intracellular target to achieve their therapeutic effect. Therefore, the tremendous therapeutic potential of nucleic acids depends on the success to find suitable carriers that will bring them to their target site.

APPLICATION OF MICRO- AND NANO-ELECTROMECHANICAL DEVICES TO DRUG DELIVERY

Drug delivery devices based on micro- and nano-electromechanical systems (MEMS or NEMS) offer opportunities to address unmet medical needs related to dosing. Such devices should be considered when conventional dosing methods perform sub-optimally in terms of safety, efficacy, pain, or convenience. In addition, applications of these technologies may create totally new drug delivery paradigms. MEMS technologies may create new therapies with existing molecular entities. This chapter addresses progress and prospects for combination product implementations of MEMS- and NEMS-based polymeric and electromechanical delivery devices. Factors limiting the capabilities and convenience of conventional drug administration may include long-term treatment, a narrow therapeutic window, a complex dosing schedule, combination therapy, an individualized or emergency-based dosing regimen, and labile active ingredient.

These limitations are being countered as new approaches emerge for developing drug and medical device combinations that can protect labile active ingredients, precisely control drug release kinetics (timing and amount), deliver multiple doses, eliminate frequent injection, and/or modulate release using integrated sensor feedback. Innovative delivery devices have the capability to completely control drug release: doses may be administered in pulses or continuously for periods of months to years, or doses may be stored in a device pending immediate need for emergency administration. Technologies contributing to advanced drug delivery system design include MEMS or NEMS, materials science, data management (gathering, using, and communicating data), and biological science. Nanotechnology encompasses an additional group of emerging technologies that are primarily defined by the nanometre scale to which they are applicable.

Proteins, nucleic acids, and other biomolecules are also in this size range. Micro- and nanotechnologies overlap with regard to some methods and applications. However, the ability to control materials, surfaces, and structures at the nanometre scale can provide distinctive solutions as a consequence of properties unique to the nanometre scale. New technologies are only useful if they can be commercialized, and drug delivery applications cannot be commercialized without a regulatory environment sufficiently adaptable to support marketing approval of innovative products. In 2002, the Food and Drug Administration (FDA) created the Office of Combination Products (OCP) to provide an appropriate regulatory framework for products that do not fit the established categories of drugs, devices, and biologics. A combination product may include a drug and a device; a biological product and a device; a drug and a biological product; or a drug, device, and a biological product. The combination products surveyed in this review are composed of two or more elements. At least one element will be a drug or biologic, and at least one element will be a drug delivery device constructed using MEMS- or NEMS-based technologies.

Converging Technologies using MEMS and NEMS

MEMS enables the manufacture of small devices using microfabrication techniques similar to the ones that are used to create silicon computer chips. MEMS technology has been used to construct microreservoirs, micropumps, nanoporous membranes, nanoparticles, valves, sensors, and other structures using biocompatible materials appropriate for drug administration. Using NEMS, complex mechanical nanostructures can be built with lateral dimensions as small as tens of nanometres. By incorporating transducers, control and measurement functions can be built into these systems. Innovative nanometre-scale materials and structures include quantum dots, nanowires, and nanotubes. MEMS- or NEMS-based devices are fabricated by adapting techniques developed for the electronics industry to integrate complex programmable and structural elements on a substrate. Originally the substrate was silicon, but (especially for biologically and medically oriented applications) the category is no longer limited to silicon-based devices. Also, initially the object was to integrate electrical and mechanical features that could be controlled by programmable circuitry. In addition to programming-based control, however, specific properties of composition and geometry may be enlisted to determine amount of drug release, timing, and rates.

Structures have been constructed using MEMS technologies with arrays of uniform channels as narrow as 7 nm. Protein diffusion kinetics under sink (dilute) conditions across nanoporous membranes are non-Fickian as the nanopore width approaches the hydrodynamic diameter of the solute, and Fickian the name originated from Fick's laws of diffusion which help to describe the non convective mass flux of a solute in a porous medium and Fick's equation prescribes a linear relationship between the dispersive mass flux and the concentration gradient. It is frequently used in MEMS. At greater pore widths. A 13-nm nanopore membrane was loaded with radiolabelled bovine serum albumin and implanted in rats to test its suitability for drug delivery. Slow release of protein was demonstrated, indicating that devices constructed in this fashion could be designed with high loading capacity and could deliver proteins with zero-order release kinetics. Performance and reliability impairment of MEMS- and NEMS-based devices due to tribological (friction and wear-related) problems need to be better understood, because wear properties may limit practical implementation of micro- and nanoscale devices for many applications. Improvements in atomic force microscopy methods may improve characterization of tribological properties of materials in micro- and nanoscale systems where relatively high sliding velocities (up to 10 mm/s) occur.

Alternative materials for constructing MEMS- and NEMS-based devices are being actively investigated to improve reliability and design flexibility, and to decrease cost. Polymers have superior properties to silicon for BioMEMS applications with regard to cost and versatility of physical properties. To provide more design flexibility, biocompatible polymers like polymethylmethacrylate (PMMA) and polydimethylsiloxane (PDMS) are being investigated as alternative materials to silicon. Data management, device control, and communication information science and wireless communication technology play multifaceted roles in expanding the potential utility of MEMS-based drug delivery systems. Commercial software and hardware are readily available for gathering, transmitting, manipulating, storing, retrieving, and classifying recorded information. Data can be transmitted using wireless communication from a biomonitor to a central system or to a drug delivery device as part of a feedback loop. The data itself may serve as a passive log, an alert system for patients and health care providers, or a basis for controlling drug release from a delivery device. Autonomous self-monitoring delivery systems require algorithms for translating monitor data into the control commands for timing and amount of drug to release.

The continuous glucose monitoring field, in particular, has developed a number of algorithms to translate raw data from glucose sensors into alerts for impending hypoglycaemia, although the ultimate goal of an implantable closed-loop system with automatic control of insulin delivery has not been realized. The convergence of information technology and biological (and medically related) applications has become known as "Bio-IT". Data interpretation may be complex, involving interrelated environmental variables, noise and drift, possibly requiring an artificial neural network (ANN) to generate control algorithms. ANNs are mathematical constructs that contain interconnected processing elements ("neurons"), schematically similar to biological neural connections. The neurons accept multiple inputs, apply a weighting system to the summed inputs, and produce an output signal to another neuron.

The interunit connections are optimized during the training process until the error in predictions is minimized and a targeted accuracy level is achieved. The ANN can be fed with new input information post-training to make decisions or perform actions. ANNs have a number of properties and capabilities that are important for critical, computation-intensive applications: parallel processing, fault tolerance, self-organization, generalization ability, and continuous adaptivity. ANNs are being applied to a wide range of data-intensive medical applications, including evaluation of physiological data, epidemiological phenomena, medical image analysis, and monitoring the effectiveness of treatment regimens. Pharmaceutical research has used ANNs for tasks such as evaluation of analytical data, drug design, dosage form design (formulation and delivery), and pharmacokinetic/pharmacodynamic modelling.

Biological Sciences

Advances in biological sciences have produced many new drugs and biological molecules that are candidates for administration via drug delivery devices due to factors such as stability requirements and high potency. Manufacturing operations to produce biological molecules (recombinant DNA, cell fusion, and new bioprocessing techniques) have not changed greatly in the last 5 years, but systems biology approaches to understanding links between the genome (via the Human Genome Project), proteins ("proteomics") and metabolites ("metabolomics") have blossomed into a set of fields that has the potential to greatly enrich discovery of therapeutic drugs and biomolecules.

REGULATORY DIMENSIONS

It is critical to understand the fundamentals of current regulatory procedures because of the complexity of obtaining combination product approval. Additionally, relevant procedures and guidance have been evolving extensively in the last several years. Although no MEMS- or NEMS-based combination products have been approved yet. MEMS- and NEMS-based systems will primarily be differentiated on the basis of whether micro- or nanoscale features are the most critical contributors of functionality. Although the FDA recognizes the need to specifically address requirements for product manufactured using nanotechnology-based processes, there are currently no testing or safety evaluation requirements specific to nanotechnology products, and the FDA does not anticipate any new guidance documents regarding nanomaterials in the near future. Many aspects of these products that concern quality and efficacy are already encompassed by existing guidances for more conventional products. The FDA–OCP will coordinate the regulatory framework for nanotechnology products and will designate the FDA center responsible for evaluating the application, with consultation from other centres.

As an increasing number of innovative drug delivery products were presented to the FDA for approval, it became clear that the division of products into drugs, biologics, and devices was not an adequate system for evaluation. The OCP assigns a lead center for premarket review and regulation of combination products based on the primary mode of action and coordinates a wide range of administrative functions related to combination product review and approval. The FDA provides a device definition that distinguishes between devices and drugs or biologics.

A device provides diagnostic or therapeutic benefit and "does not achieve any of its primary intended purposes through chemical action within or on the body of man or other animals and which is not dependent upon being metabolized for the achievement of any of its primary intended purposes". If the primary product use involves chemical action or metabolism of the product, this is the key consideration for identifying a product as a drug. A drug is intended for use "in diagnosis, cure, mitigation, treatment, or prevention of disease, or an article (other than food) intended to affect the structure or function of the body."

There are some regulatory hurdles specific to combination products that can add uncertainty, lengthen the preparation for submitting an application to the FDA, or delay the approval process. In the United States, the OCP assigns the FDA center that has "primary jurisdiction" and the type of application [premarket approval (PMA), new drug application (NDA), or biologics license application (BLA)] that must be prepared, based on the primary mode of action. From an organizational skills standpoint, drug and device companies have very different sets of expertise. Typically, a device company will have difficulty preparing a successful NDA or BLA. Consequently, assignment of a combination product to a regulatory centre unfamiliar to the applicant can cause additional applicant uncertainty and delay in approval. Another important consideration involves judging when the formulation, dosing amount, dosing frequency, or delivery mechanism sufficiently differ from an approved NDA, that a new NDA is required. Cross-labelling (correspondence between device labelling and drug labelling) must also be resolved. The most challenging part of the submission will often be the data interpretation of *in vivo* and clinical studies: if a case cannot be made that the pharmacokinetic/pharmacodynamic data, toxicology, safety, dosing, and efficacy are sufficiently similar for the active ingredient in an approved product and the combination product, additional clinical studies may be required.

A combination product application also requires technical details for the interaction of each component as well as data on the components themselves, and the amount and type of data must be considered on a case-by-case basis. Quality control groups need to provide additional testing specific to the drug/device combination. As an example, for drug-coated stents, testing should measure the impact of the polymer coating on drug efficacy. The drug-release rate and factors that affect the release from the stent as well as the optimal *in vivo* release profile should be determined. Localized effects should be identified, and *in vitro* test methods specific to the combination product should be designed. Stability testing must include the impact of the product components on each other. A device must be proven safe and effective for each drug used with it even if the device design and operation remain the same because the system performance may depend on how the device (for instance, a pump) and drug interact. Similarly, drug chemistry and packaging compatibility must be verified for each new combination of products.

IMPLANTABLE DEVICES

Reservoirs for Controlled Release

A biodegradable polymer chip version of an implantable multireservoir drug delivery device incorporates an array of reservoirs capped with resorbable membranes that may differ from other membranes in the array by thickness or chemical composition. The interior of each reservoir contains drug formulation(s). An advantage of biodegradable polymer-based systems compared to microchip-based systems is the elimination of a requirement for a second surgery to remove the device. In addition, the lack of electronics reduces any size restrictions in terms of device manufacture. Such systems are simpler but will not deliver reservoir contents with as much precision as the analogous microelectronic devices.

These polymeric devices and the reservoirs are formed by compression-molding polylactic acid (PLA). Individual membrane recipes are prepared using various ratios of lactic acid/glycolic acid and different molecular weight polymers to control release of reservoir contents from the devices. Prototype devices were approximately 11.9 mm in diameter, 480–560 mm thick, and contained 36 individual 120–130-nL reservoirs. Poly(lactide-co-glycolide) (PLGA) polymer membranes representing a molecular weight range of 4400 to 64,000 Da were used to control the release rates of the polymers namely human growth hormone (HGH), dextran, and heparin. The activation of reservoirs in sequence was illustrated by *in vitro* release of HGH, dextran, and heparin. Similar results are obtained *in vivo*. The device has been shown to be biocompatible. ChipRx® (Lexington, KY, USA) has proposed an implantable, single-reservoir device that, in theory, could be adjusted to deliver drugs with targeted pharmacokinetics and bioavailability. The release mechanism employs polymeric "artificial muscles" that ring holes that have micrometre-sized diameter holes and that open to release drug. The polymer ring expands or contracts in response to an electrical signal transmitted through a conducting polymer that contacts a swellable hydrogel. Development work on this "smart pill" has been suspended by ChipRx®, however.

Stents

Drug-eluting stents [i.e., Taxusi® from Boston Scientific (Natick, MA, USA)] are among the most widely known combination products. Advances in technology for cardiovascular stents have guided products through several generations. Micromachining technology allowed bare metal stents to be manufactured that had the physical capability of propping open occluded vessels. Coating the stent with a drug-containing polymer resulted in combination products featuring localized drug release capability in addition to the mechanical action of the stent. Next generation drug-eluting stents incorporate reservoir-based drug containment on the stent surface, with release properties determined by polymer composition and layer thickness and achieves flexible and controllable pharmacokinetic profiles of paclitaxel release through layered polymer/drug inlay stent technology. Programmable, complex chemotherapy using this approach may be feasible for the treatment of cardiovascular disease.

Figure 4.1 illustrates how several configurations of drug and polymer barrier can be layered in the stent reservoirs to control amount and timing of drug release. Bare metal stents may also be embellished with microprobes that allow delivery of antirestenosis drugs or biologicals. This stent design has been tested *in vivo* in rabbit femoral arteries. The microprobes penetrate the atherosclerotic plaque

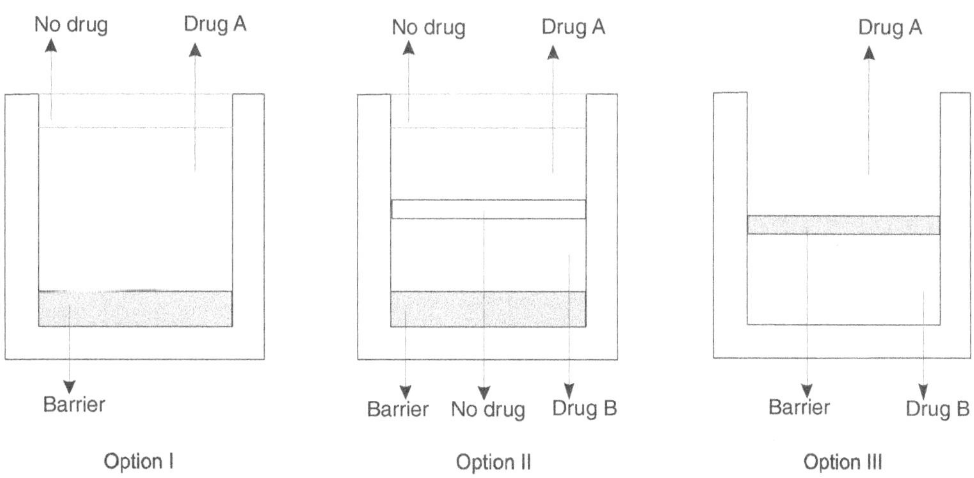

Figure 4.1 Options in Filling Reservoir with Drug

and the internal elastic lamina, reducing the diffusion barrier layer for delivery of genes or drugs. The device uses controlled balloon expansion to cause the microprobes to pivot upward during inflation, lifting them into position. The design specifies the microprobe penetration depth, so the stent can be tailored to vessel sizes. The outer surface of the device is a nanoporous layer that controls active agent delivery. Addition of communication capabilities provides another way to increase stent functionality. A micromachined stent has been developed that serves as an antenna for wireless monitoring of implantable microsensors. The stent expands into an inductive coil after implantation, after links in the structure break and alter the electrical characteristics.

Enteral (Mucosal) Delivery

The mucosal route offers many advantages for drug delivery, especially for peptides and proteins. Drug bioavailability is improved due to avoidance of degradation in the gastrointestinal tract and hepatic first-pass metabolism. However, a short drug residence time, the presence of enzymes, and a limited permeability of the epithelial barrier are the main drawbacks of mucosal administration. Microfabrication technology, combined with appropriate surface chemistry, may permit the localized dosing and unidirectional release of therapeutic agents. Poor oral bioavailability of polypeptide drugs can be alleviated by fabricating particles of 0.1–1.0 mm diameter using MEMS technology. Reservoirs in these particles contain controlled-release drug formulations (or drug is sealed in the reservoir with an erodible polymer cap). A mucoadhesive agent on the same face as the opening of the drug reservoir orients drug release toward the mucosal surface. These particles are enteric coated, allowing passage through the stomach before releasing the drug. The particle may be fabricated by a "top-down" approach that combines thin film deposition methods, photolithography, photoablation, and etching techniques. Microfabrication techniques and surface modification have been combined to create well-controlled, biologically specific drug delivery modules.

The delivery module adheres to the gastrointestinal tract while the drug is released from the same face. Reservoir-containing silicon microdevices have been modified to enhance bioadhesion *in vitro* with a surface chemistry that binds lectin via avidin–biotin interactions. Lectins are a group of proteins that can bind specific cell types in the gastrointestinal tract. This strategy could improve oral bioavailability of therapeutic biopolymers such as peptides, proteins, and oligonucleotides. In a related study, a similar approach involved lectin attachment to PMMA microdevices. Aminolysis was used to modify the PMMA to yield accessible amino groups on the PMMA surface, followed by reaction with an active ester to covalently bind avidin molecules to the surface of the particles. Subsequent incubation in a biotinylated lectin solution produced lectin-modified microdevices that were tested *in vitro* for bioadhesion. The geometry of the isotropic reservoir maximizes epithelial-specific drug delivery, while minimizing drug dilution and metabolism in the gastrointestinal tract.

PROSPECTS

Advances in many fields are converging to make commercialization of advanced drug delivery concepts possible. MEMS and NEMS, materials science, information technology, ANNs, wireless communication, and systems biology can all contribute to design of integrated therapeutic systems that have the potential to significantly improve the quality of pharmaceuticals-based medical care. Table 4.1 summarizes examples of commercially available drug delivery devices, devices that are in clinical trials, and devices that are at the research or preclinical testing stage. Typically, a primary motivator for improved drug delivery systems is the avoidance of repeated parenteral administration. Due to the complexity and cost of device-based methods of administration, if a therapy can be accomplished by oral, pulmonary, or other nonparenteral routes, it is unlikely that sufficient advantage will be gained by introduction of an advanced delivery device. Additional requirements for a commercial device include highly potent active ingredients for implantable devices and stable formulations for long-term administration. For instances where a closed-loop system would be highly desirable, the use of an implanted sensor with drug administered by an external pump may offer significant advantage (e.g. the insulin pump combined with a glucose monitor described below). NEMS-based devices are still at an early stage of testing. For implantable devices, nanotechnology for drug delivery is more likely to provide incremental rather than major advances because drug potency limits the minimum size of an implant for chronic administration.

Added value may be obtained with regard to increasingly personalized medicine, customized delivery, and implementation of feedback loops between biosensors and control of drug dosing (amount and timing) by applying the evolving tools of information technology to drug delivery. Cardiac pacemakers and defibrillators are examples of commercially available systems that prove the practicality of this concept. Clinicians and other health care providers, and maybe each individual patient, could access real-time medical status and order

TABLE 4.1 OVERVIEW OF DEVELOPMENT OF MEMS- AND NEMS-BASED DRUG DELIVERY DEVICES

Period	Drug delivery devices (under clinical trials)
Present	
(Commercial products; not necessarily MEMS- or NEMS-based but lay the foundation for MEMS- or NEMS-based drug delivery)	• Reservoir-based drug-eluting stents (Conor Medsystems) • Transdermal patches: iontophoresis devices • Microosmotic pumps • Closed loop system for insulin administration: insulin pump controlled by glucose sensor
Near-term	
(In clinical trials; 1–10 years to market)	• Microneedle-based transdermal delivery
Mid- to long-term	
(Research or preclinical; 5–25 years to market)	• Microchip technology: electrochemical drug release • Microchip technology: electrothermal drug release • Biodegradable polymer chip technology • Microchip technology: "smart polymer" drug release • Selective membrane technologies: for drug delivery; various others in research • Stent technologies: microprobe-based stent technology; stents as antennas for monitoring implanted sensors • Various microneedle devices in research • Transdermal drug delivery/biosensor feedback loop • Transcutaneous drug delivery/biosensor feedback loop • Micropump for drug delivery and many other micropumps in research • Various micro- and nanotechnology implementations of controlled drug release; mucoadhesive particles for oral administration

intervention via controlled dosing. Artificial intelligence and ANN research are being used to develop rules for computer systems and software to mimic biological processes for reasoning, pattern recognition, and processing of sensor data. It may be possible to incorporate ANNs into MEMS-based devices, or such decision-making capabilities could reside external to the device but accessible through wireless communication.

Wireless communication allows flexibility in integrated device design, as a device can be physically separated into modules without sacrificing system capabilities. A system providing complete control and feedback, for example, could be implanted as a biosensor unit and a drug reservoir unit. The units could communicate to regulate drug release and could also receive additional instruction from an external agent, as well as send data to an external monitor for use by patient or physician.

A large population of type 1 and type 2 diabetics could greatly benefit from an "artificial pancreas," the most commonly cited example of an integrated medical system. Better control of insulin administration by taking real-time glucose-monitoring data into consideration would significantly decrease diabetic complications. To construct a commercializable system, four components must be integrated: sampling, glucose sensing, mathematical models and related algorithms to calculate insulin doses, and the insulin delivery system. Currently, the main hurdles to realizing this goal are continuous glucose monitoring and the necessary control algorithms. Medtronic MiniMed, Inc. (Northridge, CA, USA) and Becton Dickinson (Franklin Lakes, NJ, USA) market a device/device-combination product for diabetics that "integrates a glucose meter and an insulin pump with a dose calculator into one device as a step toward development of a fully automated glucose monitoring and insulin delivery system." In general, drug/biosensor/delivery device combinations could be designed to respond to specific signals or circumstances, providing more responsiveness to dosing adjustments than is possible with traditionally administered drugs.

Some multidisciplinary advances will support advanced drug delivery. Recent regulatory initiatives, such as formation of the FDA-OCP and the FDA NanoTechnology Interest Group, have helped clarify the pathway for marketing approval of these innovative products. The combination of new technology and flexible regulatory guidance promises to foster further development of innovative drug delivery combination products for the foreseeable future.

PHOTODYNAMIC THERAPY IN TARGETED DRUG ADMINISTRATION

Photodynamic therapy (PDT) is defined as the site-directed generation of cytotoxic effects upon targeted light administration to activate photosensitive chemical compounds. Light activation of chemicals or dyes for photo-killing was first introduced in the beginning of the 20th century and is gaining recognition as an effective modality against cancerous as well as noncancerous diseases. To produce cytotoxic effects, PDT requires three components to be present simultaneously: light, oxygen and a photosensitizer (PS). Various PS have been approved for clinical use, including porfimer sodium (Photofrin) for the treatment of lung and digestive tract cancers and Barrett's esophagus; aminolevulinic acid (ALA) for the treatment of actinic keratoses; and temoporfin (Foscan), which is approved in the European Union and Japan for the treatment of head and neck cancers. In 1993, Photofrin was initially clinically approved for the treatment of bladder cancer, and is currently the most commonly used PS. However, Photofrin has poor selectivity between tumour and normal tissues, and due to its long clearance time, produces skin photosensitivity that can last for several weeks following treatment. Second generation PS such as verteporfin (Visudyne), which is approved for the treatment of age-related macular degeneration (AMD), have improved tumour selectivity and a quicker clearance, resulting in less prolonged skin phototoxicity.

In addition to a somewhat preferential accumulation of the PS in the tumour vs in normal tissue, selectivity in PDT is achieved by precisely directing light irradiation to the tumour. While the current regimens used for PDT may be adequate for palliation, improved treatment responses (and perhaps even a targeting modality) may be necessary for more complex anatomical sites with localized but intricate three-dimensional structures, such as abdominal tumours (ovarian cancer) or tumours in the thoracic cavity (lung cancer).

COMBINATION THERAPY

Photodynamic therapy is gaining recognition as a treatment option for many cancerous as well as non-neoplastic pathologies, and the clinical results are promising. However, for PDT to evolve as a first-line curative modality, both our understanding of PDT mechanisms and the designing of improved treatments based on this understanding promise to provide an enhanced therapeutic response.

Combination regimens comprising PDT and a secondary treatment can be designed to increase the effectiveness of PDT by either: (i) reducing potentially detrimental molecular responses triggered by surviving tumour cells following PDT, or (ii) increasing the susceptibility of tumour cells to PDT. In Figure 4.2, "Approach (a)" is based on prosurvival molecular responses such as angiogenesis and inflammation, which are elicited by PDT. Another important factor governing the cytotoxic effects of PDT is hypoxia; PDT can cause hypoxia by oxygen consumption and by vasculature damage. Hypoxia is a major stimulus for angiogenesis, which is mediated by the vascular endothelial growth factor (VEGF), a potent angiogenic molecule.

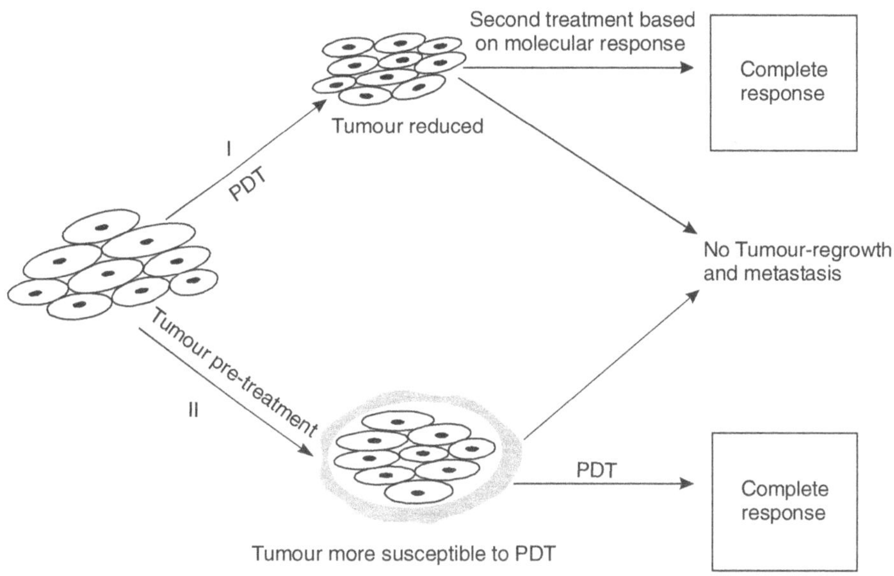

FIGURE 4.2 MOLECULAR RESPONSES OF TUMOUR TREATMENT BY PDT

TARGETING SPECIFIC CELLULAR FUNCTION-SENSITIVE LINKAGES

Specificity can be improved by not only targeting a PS to a specific site of action, but also by creating an active form of a PS from an inactive (quenched) form by using specific cellular functions (such as enzymes) at the site of action. In this case, an inactive PS is administered such that it is activated (and produces cytotoxic effects) only at the site of the lesion. This method has been utilized for both cancer and antimicrobial PDT. Funovics *et al.* described the use of various imaging probes designed specifically for locating proteases to be used as targets in the detection of tumours and other diseases. The underlying principle is that in a fluorescence resonance energy transfer (FRET) between the donor and the acceptor, the absence of proteases results in the quenched fluorescence for the donor; however, in the presence of proteases, the substrate protein is cleaved, releasing the FRET interaction between the donor–acceptor fluorophore, which results in a four-fold increase in the fluorescence signal for an initially quenched molecular dye. A large number of fluorescent probes have been described in detail for proteolysis as a tool in drug discovery.

Although it is difficult to specifically demarcate between cancerous cells and noncancerous cell markers, functional sensitive linkages have been significantly beneficial for picking specific markers for infectious diseases such as microbial infections, where enzymatic expressions are clearly differentiated from the host environment. The enzymatic compositions of bacterial and mammalian cells are significantly different, offering an opportunity for capitalizing the difference for achieving high selectivity against bacterial cells.

ENHANCEMENT OF PDT BY THE PS CONJUGATION TO CARRIER MOLECULES

Synthetic Peptides

A recent study on enhancing the selectivity of PDT describes in detail the applications of synthetic peptides and polymeric compositions (e.g. polylysine, polyarginine) as PS carriers for applications in tumour targeting and antimicrobial therapy. Peptides have advantages over other carrier molecules due to their small size, relative ease of synthesis and high affinity for binding to receptors. A range of peptide sequences have been used successfully to direct PS to target tissues which express molecules such as gonadotropin-releasing hormone, angiogenic factors, VEGF receptor-2 and neuropilin-1 (NRP-1) recombinant chimeric protein using VEGF receptor-specific heptapeptide. Synthetic peptides are also used as a nuclear localization signal (NLS) as the nucleus is the most sensitive site in a cell, and the proximity of the PS to the nucleus results in enhanced cytotoxicity. NLS peptides have also been coupled to other carrier systems, such as LDLs, to provide a potentially selective entry pathway into malignant cells that overexpress the LDL receptor, or polymers that provide improved targeting of tumour sites. These peptides need to be chemically modified for stability and efficiency, and common strategies include using pseudo amino acids and cyclic peptides. Synthetic peptides are also gaining recognition as specific tumour markers for diagnosis and have been attached to fluorescent dyes or NP for the selective imaging of cancer cells. For example, a recent study showed a disease-specific library-derived fluorescent probe for the early detection of cancer. The researchers used a phage clone (G1) displaying the peptide sequence IAGLATPGWSHWLAL, fluorescently labelled with the near-infrared fluorophore AlexaFluor 680, and evaluated its ability to bind and target PC-3 prostate carcinomas. The fluorescently labelled phage clone (G1) had a tumour-to-muscle ratio of approximately 30 to 1 and prostate tumours (PC-3) are readily detectable by optical imaging methods.

Polymers

Biodegradable polymers are extensively used for drug delivery, and when conjugated to a PS, are known to increase PDT specificity and/or uptake in tumours or other pathological lesions. These polymers as carriers alter the pharmacokinetics, the biodistribution (with less skin photosensitivity) and decrease the phototoxicity to normal tissue. Polymers can be used as micelles or encapsulating agents to solubilize poorly soluble PDT agents such as mesotetraphenylporphine (TPP). Encapsulation of TPP into polyethylene glycol/phosphatidylethanolamine conjugate (PEG-PE)-based micelles and immunomicelles (bearing an anticancer monoclonal 2C5 antibody) resulted in significantly improved anticancer effects of the drug in murine and human cancer cells. The most abundantly used biodegradable polymer to enhance the physical properties of PS is PEG. Compared with the nonpegylated conjugate, the attachment of PEG (pegylation) to a poly-L-lysine–chlorin e6 conjugate increased the relative phototoxicity in an ovarian cancer cell line (OVCAR-5), while reducing it in a macrophage cell line. This suggests that the pegylation of a polymer–PS conjugate improves tumour targeting. Considering these advantages, polymers are extensively utilized in dual modes of targeting the cancer cells.

Composite Targeting

To obtain full potential of PDT, strategies have been explored in which more than one carrier moiety was conjugated with PS for the selective targeting of cells and tissues. Considering the complex nature and architecture of tumor tissues, it is essential to design a carrier system that will maximize the cytotoxic effect on tumour cells with a minimum effect on healthy cells. These studies provide a platform for the design of optimal PDT delivery systems; NP is an example of such multimodal or composite carriers that can tag a PS with more than one targeting moiety.

Nanoparticles are at the leading edge of the rapidly developing field of material science in nanotechnology, which influences both diagnostics and therapeutics.

NP range in size from 1-1000 nm, and have beneficial properties such as biodegradability, large surface areas, high PS loading and ease of functional modification. The variety of NP constructs is increasing rapidly and covering all of them would be beyond the scope of this book. Therefore, we have focused on the emerging use of NP on PS encapsulation for PDT. Figure 4.3 shows a pictorial representation of a typical NP application for PDT. The PS is encapsulated within

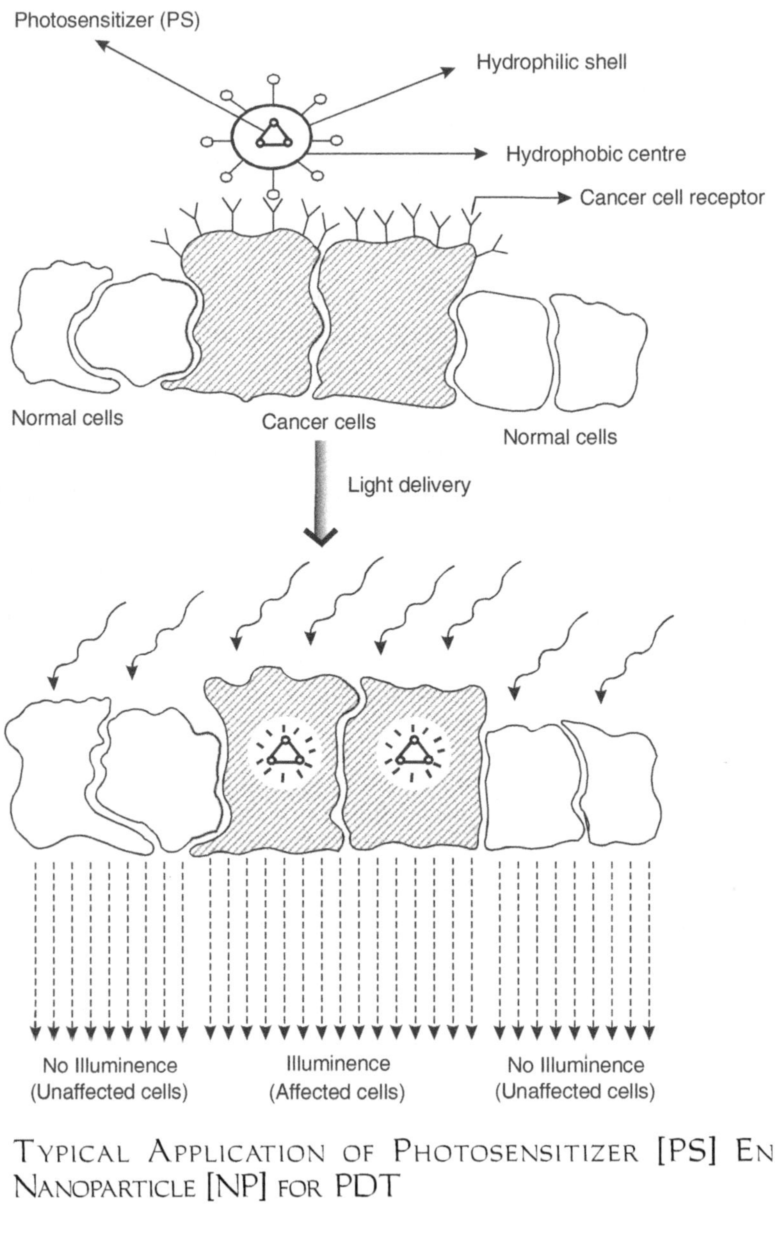

FIGURE 4.3 TYPICAL APPLICATION OF PHOTOSENSITIZER [PS] ENCAPSULATED NANOPARTICLE [NP] FOR PDT

the hydrophobic core of the NP. The outer layer, made up of hydrophilic PEG, is derivatized with a targeting moiety that selectively binds to tumour cell-surface motifs. Upon irradiation, the tumour cells are destroyed without affecting the surrounding normal tissue. Various polymers have been employed in the generation of NP to achieve biocompatibility, coupled with ease of PS conjugation or encapsulation, and modifiable surface groups. The most common NP for embedding PS are the polymers of polylactic acid (PLA) or polylactic-co-glycolic acid (PLGA).

There are two broad strategies for the enhancement of PDT treatment response. These involve: 1) mechanism-based combination treatments in which PDT and a second modality can be designed to either increase the susceptibility of tumour cells to PDT or nullify the treatment-outcome-mitigating molecular responses triggered by PDT of tumours, and 2) some of the more recent approaches of PS targeting, either by specific cellular function-sensitive linkages or via conjugation to macromolecules. PDT outcomes can be improved by the development of new combination regimens, by the chemical synthesis of conjugates that target cellular functions exclusively and by specific carrier systems that are designed to target tumour cells in more than one way. Each of the strategies comes with significant challenges. For instance, rationally designed combination treatments may require extensive mechanistic studies and may be patient-specific. The best results in this case may be obtained with individualized regimens, which would be a complicated undertaking. In the case of targeted PDT, potential challenges include the complexity of new chemistry, the purification of conjugates and the low PS-loading onto carrier molecules. The development of specific cellular function-sensitive linkages for cancer targeting also faces a difficulty in that it may be easier to develop functional targeting approaches for antimicrobial PDT, as the molecular expression of enzymes is more clearly distinguished from the surrounding host environment when compared with the functional differences between tumour cells and normal cells. Further exploration of strategies that target the PS to tumour cells and mechanism-based combination therapies that enhance the treatment outcomes of PDT may lead to new and improved treatments for a variety of cancers and other pathologies, including for those diseases that are resistant to other therapies.

ADVANCES IN THE MANUFACTURING, TYPES, AND APPLICATIONS OF BIOSENSORS

In recent years, there have been significant technological advancements in the manufacturing, types, and applications of biosensors. Applications include clinical and non-clinical diagnostics for home, bio-defence, bio-remediation, environment, agriculture, and the food industry. Biosensors have progressed beyond the detection of biological threats such as anthrax and are finding use in a number of non-biological applications. Emerging biosensor technologies such as lab-on-a-chip have revolutionized the integration approaches for a very flexible, innovative, and user-friendly platform.

A biosensor is a device that detects, records, and transmits information regarding a physiological change or the presence of various chemical or biological materials. These biological materials include enzymes, tissues, microorganisms, antibodies, cell receptors, and biologically derived materials. Biosensors also have a mimicking component due to intimate contact with a physio-chemical transducer or transducing microsystems in the environment. In other words, biosensors are analytical devices incorporating biological materials operating via a biorecognition process, which can be due to their bio-affinity or bio-metabolism. The results of this recognition process will then be converted through a transduction mechanism using transducers, which are the components that convert a biochemical signal into a quantifiable electrical signal. Consequently, the output of this transduction mechanism can be an optical, electrochemical, piezoelectric, thermometric, magnetic, or even calorimetric signal that can be used to quantify the analyte presented. Figure 4.4 represents a schematic representation of a biosensor.

In Figure 4.5, the various components of a biosensor are described. Technologically, a biosensor is a probe that integrates a biological component, such as a whole bacterium or a biological product (e.g. an enzyme or antibody) with an electronic component to yield a measurable signal. Biosensors, which come in a large variety of sizes and shapes, are used to monitor changes in environmental conditions. They can detect and measure concentrations of specific bacteria or hazardous chemicals as well as measure acidity levels (pH). Biosensors have found applications in a variety of scenarios. Due to their simplicity, high sensitivity, and potential for real-time and on-site analysis, biosensors have been widely applied in various fields in medical diagnostics, environmental monitoring and genetics, health care, patient management, food processing, and defence including industrial processes, clinical detection, and environmental control. Developing biosensors that can be scaled down to a small footprint with reproducible performance and accurate technical standards is a critical factor for manufacturing. In addition, the cost of manufacturing coupled with factors such as the need for sterilization, repeatability, and time budget for data acquisition and analyses need to be considered.

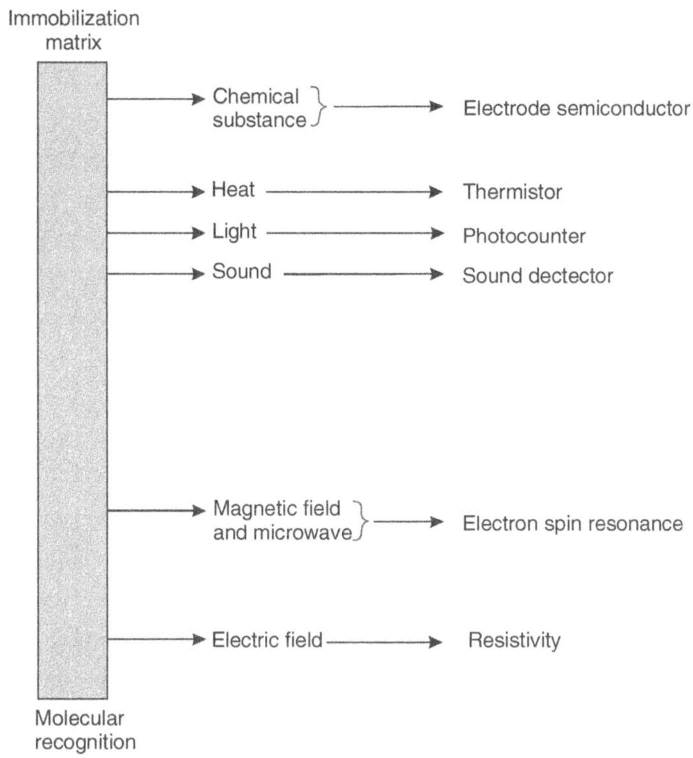

FIGURE 4.4 A SCHEMATIC REPRESENTATION OF BIOSENSOR

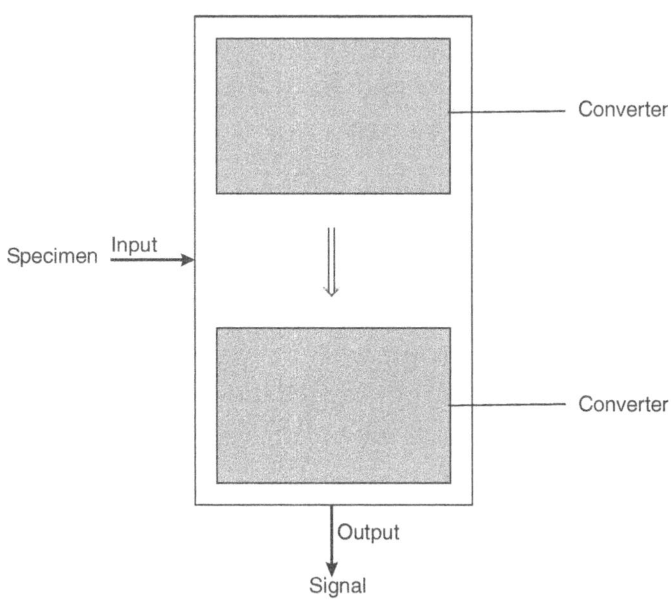

FIGURE 4.5 THE VARIOUS COMPONENTS OF A BIOSENSOR

Biosensor Types

There are various types of biosensors based on the principle of detection, including optical, piezoelectric, electrochemical, and thermometric. Optical biosensors, such as Farby–Perot, detect changes in absorbance or fluorescence of an appropriate indicator and changes in the refractive index. Piezoelectric sensors are based on an alternating potential and produce a standing wave in the crystal at a characteristic frequency. This frequency is highly sensitive to the surface properties of the crystal such that, if a crystal is coated with a biological recognition element, the binding of the target analyte to a receptor will produce a change in the resonant frequency. Electrochemical biosensors are based on enzymatic catalysis of a reaction, which produces ions. The sensor substrate contains three electrodes: a reference electrode, an active electrode, and a sink electrode. A counter electrode can also be present as an ion source. The target analyte participates in the reaction that takes place on the active electrode surface, and the ions produced create a potential that is subtracted from that of the reference electrode to yield a signal. Electrochemical biosensors are usually based on potentiometry and amperometry. Thermometric biosensors are constructed by combining enzymes with temperature sensors. When the analyte is exposed to the enzyme, the heat of reaction of the enzyme is measured and is calibrated against the analyte concentration. The core requirements for a biosensor approach to be valuable in terms of its utility and manufacturing are identification of a receptor that binds to the target of interest; availability of a suitable biological recognition element; and potential for disposable portable detection systems as opposed to laboratory-based techniques in some situations.

Biosensor Developments

In the mid-1980s, Tuan Vo-Dinh and collaborators at the Life Sciences Division at Oak Ridge National Laboratory were looking for a way to use light to detect cancer-causing agents in groundwater. Their successful results led to the development of a series of fibre-optic-based biosensors. Screen-printing biosensors using enzymes as the bio-catalysts are a platform technology, particularly as specific catalysts or modifications are made or incorporated into the sensing elements. Ink-jet printing technology is a non-contact process that can dispense a well-defined micro-quantity of the enzyme. While the physics and the approaches of various ink-jet printing may be different, the outcomes of the ink-jet printing are similar in obtaining the well-controlled, micro-size droplets.

Bacteria can be used as biosensors to demonstrate the toxicity of a variety of environmental media including soil, sediment, and water by coupling bacteria to transducers that convert a cellular response into detectable signals. These bacterial biosensors are engineered by pairing a reporter gene that generates a signal with a contaminant-sensing component that responds to chemical or physical change, such as exposure to a specific analyte. When the biosensor is exposed to such a change, the sensing component stimulates the reporter gene through a biochemical pathway in the cell. The reporter gene then produces a measurable response such as emitting visible light, indicating the degree of chemical or physical change.

Several biosensors have been developed that indicate toxicity of any chemical or physical change; new biosensors are being developed to respond to particular analytes. Such biosensors have been developed for heavy metals and metalloids including arsenic, cadmium, mercury, and lead. The RAPTOR® is a portable automated fibre optic biosensor for detection of biological threat agents. It performs rapid (three to ten minute), fluorescent sandwich immunoassays on the surface of short polystyrene optical probes for up to four target analytes simultaneously. Over the past 3–5 years, optical biosensors have demonstrated the sensitivity required for typical drug candidates and lower molecular weight drug fragments or "needles." The sensitivity for drug needles is poorer than for larger molecular weight drugs as surface plasmon resonance (SPR) measures changes in refractive index that are directly related to the molecular weight of the binding molecule.

After more than 20 years of research and development, SPR sensitivity may be approaching theoretical limits in terms of the detection interface sensitivity. Bioreporters refer to intact, living microbial cells that have been genetically engineered to produce a measurable signal in response to a specific chemical or physical agent in their environment. Bioreporters contain two essential genetic elements, a promoter gene and a reporter gene. The promoter gene is turned on (transcribed) when the target agent is present in the cell's environment. The promoter gene in a normal bacterial cell is linked to other genes that are then likewise transcribed and then translated into proteins that help the cell in either combating or adapting to the agent to which it has been exposed. In the case of a bioreporter, these genes, or portions thereof, have been removed and replaced with a reporter gene. Consequently, turning on the promoter gene now causes the reporter gene to be turned on. Activation of the reporter gene leads to production of reporter proteins that ultimately generate some type of a detectable signal. Therefore, the presence of a signal indicates that the bioreporter has sensed a particular target agent in its environment.

Biosensor Market—Manufacturers, Potential, and Drivers

Rapid scaling down in size, high throughput, low power dissipation, ease of calibration, and low cost will be the key factors that will quicken expanded commercialization of biosensors to penetrate several untapped markets. Manufacturers have successfully integrated biosensor technology with leading-edge integrated circuit and wireless technology in high-end applications. Recently, there has been a tremendous push for molecular foundries, electronic printing and disposable sensors. This chapter clearly discusses the potentials of biosensors to overcome most of the disadvantages of conventional methods. Novel integration technologies, such as magnetic field-assisted assembly, will assist in developing methodologies for implementing schemes such as lab-on-a-chip for use as biosensors. This, coupled with self-assembled enzyme aggregates prepared from magnetic iron oxide nanoparticles, will have significant utility in biosensor applications. The total global market for biosensors and bioelectronics is expected to grow from \$ 6.96 billion in 2006 to \$ 8.2 billion in 2009, at an average annual growth rate of about 6.3% (Figure 4.6). Glucose sensors accounted for nearly all

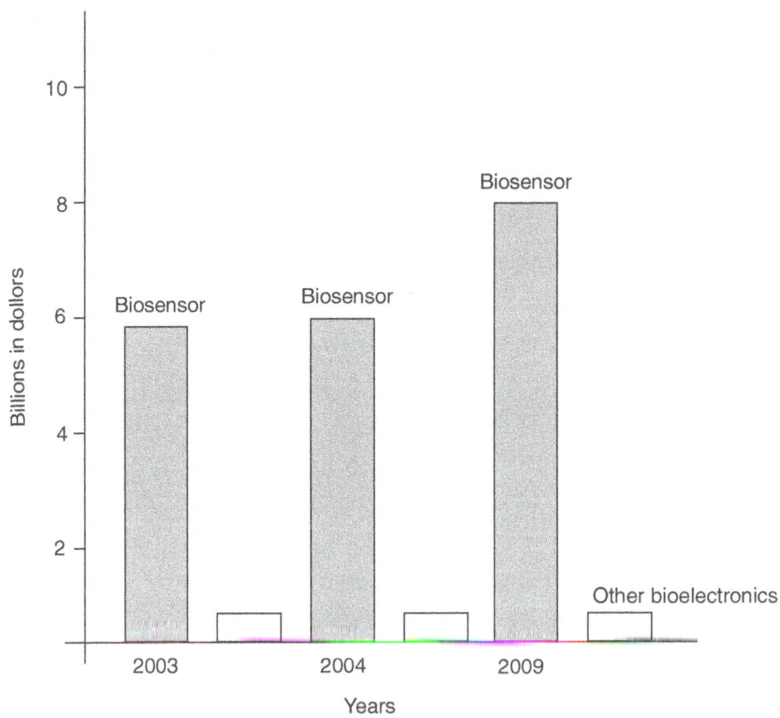

Figure 4.6 The Growth of Biosensor Industry

of the market in 2003. Sales of other bioelectronic devices are projected to increase significantly over the next five years. Biomedical and life sciences applications account for 99% of the market, with environmental monitoring and remediation applications a distant second.

Some of the key market drivers will include the following technologies:

- Detection of drug delivery, characterization of cancer cells/AIDS virus, selection of the ideal drug, and targeted drug delivery for treatment of diseases such as cancer and AIDS, frictionless nanofluidics platform that will be lossless and contamination-free, and requires transport, distribution, and mixing of nanolitre volumes.
- Radio-frequency-triggered drug delivery systems that will use plasmon-enhanced biosensor array and will require quantitative assessment of biomolecules with high accuracy, selectivity, and sensitivity.
- Radio-frequency-triggered drug delivery systems that will perform drug delivery through coated magnetic nanoparticles contained in the targeted tumour using a magnetic field.
- Biosensor devices such as the SmartPill pH.p® capsule that can measure pressure, pH, and temperature from within the entire gastrointestinal tract using miniaturized on-board sensor technology; the SmartPill pH.p® capsule assists in the evaluation of motility disorders, such as gastroparesis.
- Devices to detect biological agents such as botulinum toxins that can contaminate water and anthrax that can destroy the fundamental global infrastructure.
- Innovative multidisciplinary approaches including printed intelligence that will integrate mass manufacturing methods such as ink-like innovative material technology and functionalities created from electronics, biotechnology, chemistry, optics, optoelectronics, or their combinations.

BIOMEDICAL SENSORS AND BIOSENSORS

Sensors in Modern Medicine

A convergence of factors is now resulting in the rapid development of sensors for biomedical use. These factors include:

i. Increasing knowledge of the physics, chemistry and biochemistry of physiology
ii. Pressure for reduction in the cost of delivering medical care through more efficient treatment and shorter hospital stays
iii. Steadily decreasing costs of microprocessor technology for data acquisition, analysis and display
iv. Reduction in the size of the sensors due to silicon microfabrication and fibre optic technology
v. Rapidly advancing sensor technology in nonbiomedical fields and for *in vitro* use for clinical chemistry
vi. Advances in biomaterials

Advances in computer technology have reached the point where the control of devices and processes is often limited only by the ability to provide reliable information to the computer. Furthermore, the technologies developed by the microprocessor and fibre-optics industries are now spawning a new generation of sensors. While there has been ready adoption of new sensor technologies for there has been for *in vitro* measurements, *in vivo* use of such sensors to clinical practice is lagging far behind. This is due to serious performance deficits of sensors operation *in vivo*, as well as the same regulatory factors that apply to any use of devices and material *in vivo*. Sensors are as subject to the same problem of biocompatibility as are any other type of *in vivo* devices.

Physical vs Chemical Sensors

Biomedical sensors fall into two general categories: physical and chemical. Physical parameters of biomedical importance include pressure, volume, flow, electrical potential, and temperature, of which pressure, temperature and flow are generally the most clinically significant and lend themselves to the use of small *in vivo* sensors. Chemical sensing generally involves the determination of the concentration of a chemical species in a volume of gas, liquid or tissue. The species can vary in size from the H^+ ion to a live pathogen, and when the species is complex, an interaction with another biological entity is required to recognize it. When such entity is employed, the sensor is considered a biosensor. The clinical utility of monitoring compound in the body has motivated great efforts to develop biosensors. The prime target is improvement upon current method for determination of glucose concentration for treatment of diabetes. Because of the large number of diabetics worldwide and their requirement for frequent measurements of blood glucose, glucose monitoring continues to be the dominant market for enzyme-based biosensors.

Interaction of the Sensor with its Environment

One helpful way to classify sensors is to consider the relationship between the sensor and the analyte. The more intimate the contact between the sensor and the analyte, the more complete is the information about the nature of the chemical species being measured. However, obtaining greater chemical information may involve some hazard to physical condition of the system being studied. Non-contacting sensors produce only a minimal perturbation of the sample to be monitored. In general, such measurements are limited to the use of electromagnetic radiation such as light, or sampling the gas or liquid phase near a sample. It may be even necessary to add a probe molecule to the sample to make the determination. Two examples are monitoring the temperature of a sample by its infrared emission intensity, and spectroscopic monitoring of the pH-dependence of the optical absorption of the pH-sensitive dye but chemical selectivity is difficult to achieve *in vivo*.

Contacting sensors may be either noninvasive or invasive; direct physical contact with the sample allows a rich exchange of chemical information; thus much efforts have been expended to develop practical contacting sensors for biomedical purposes. A temperature probe can be either category, but with the exception of removing or adding small quantities of heat, it does not change the environment of the sample. Few chemical sensors approach the nonperturbing nature of physical sensors. All invasive sensors damage the biological system to a certain extent, and physical damage invariably leads to at least localized chemical change.

Most contacting sensors are derived from chemical assays first developed as sample removal sensors. Although it is certainly invasive and traumatic to remove blood or tissue from a live animal, removal of some fluids such as urine and saliva can be achieved without trauma. Once removed, the fluid can be pretreated to make it less likely to adversely affect a functioning of a sensor. For example, heparin can be easily added to blood to prevent clotting in an optical measurement cell. Cells that might interfere with optical assays can also be removed prior to measurement. Toxic reagent, and probe molecules can be added at will, and samples can be fractionated to remove interfering species. The sensor and associated equipment can be of any size, be at any temperature, and use as much time as necessary to make an accurate measurement. Further, a sensor outside the body is much easier to calibrate. This approach to chemical measurement allows great flexibility in sensor design and avoids many biocompatibility problems.

Duration of Use

Two major questions in the design of any sensors are how often and for how long it is expected to be used. There are several factors to be considered:

Length of time for which monitoring is required Determination of blood glucose levels must be made for the entire lifetime of diagnosed diabetics, whereas intra-arterial blood pressure monitoring may only be needed during a few hours of surgery.

Frequency of measurements Pregnancy testing may have to occur only once a month, whereas monitoring of blood pCO_2 during an operation may have to be made several times a minute.

Reusability of the sensor Some chemical sensors contain reagents that are consumed in a single measurement. Such sensors are usually called "dip-stick sensors", such as are now found in pregnancy-testing kits. High-affinity antibodies, for example, generally bind their antigens so tightly that they cannot be reused. On the other hand, more physical and chemical sensors are capable of measuring the concentration of their analyte on a continuous basis and are therefore inherently reusable.

Lifetime of the sensor Chemical sensors have limited lifetime because of such unavoidable processes as oxidation, and although these may be extended through low-temperature storage, *in vivo* conditions are a threat to the activity of the most stable biochemical. Most sensors degrade with time, and requirement for accuracy and precision usually limit their practical lifetime, particularly when recalibration is not possible. Mechanical properties also have limited lifetime. Although a thermocouple may have a shelf life of centuries, it can be broken on its first use by excessive flexing.

Appropriateness of repeated use The need for sterility is the most important reason to avoid reuse of a reusable sensor. If it is not logistically possible or economically feasible to completely sterilize a sensor, it will only be used on a single individual, and probably only once.

Biocompatibility If the performance of a sensor degraded by continuous contact with biological tissue or if the risk to the health of the patient increases with the time in which a sensor is in place, the lifetime of the sensor may be much shorter *in vivo* than *in vitro*.

Effects of Sensor on the Body

The introduction of a sensor into the body is a traumatic event, although the degree of trauma depends on the site of placement. The gastrointestinal tract can clearly be less traumatically accessed than the pulmonary artery. Critically ill patients often must have catheters placed into their circulatory systems for monitoring blood pressure and administration of drugs, fluids and food, so that no additional trauma is caused by including a small flexible sensor in that catheter. The size, shape, flexibility and surface chemistry of the sensor probe are also important factors. The outermost materials for the chemical sensor have at least one requirement not normally placed upon structural biomaterials: they must be permeate to the analyte. Thin films of polyurethanes are permeable to at least some analytes and porous Teflon work in other case, but this issue is far resolved. Some metals and graphite can be used directly as electrodes in the body.

When short-term implantations in tissue is possible, there is initial trauma at the site of insertion. Longer-term implantation increases the risk of infection along the surface of the implant. When the site of the implantation of the probe is the circulatory system, the thrombogenicity of the probe is of paramount importance. Surface chemistry, shape, and placement within the vessel have all been shown to be of great importance in reducing the risk of embolism. Also, sensors based on chemical reactions often contain or produce toxic substances during the course of their operations, so great care must be taken to ensure that these are either not released or are at low enough levels to avoid significant risk to the patient.

Biosensors

Biosensors are sensors that use biological molecules, tissues, organisms or principles. To determine the presence or concentration of more complex biomolecules, viruses, bacteria, and parasites *in vivo*, it is necessary to borrow from nature. This definition is broad and by no means universally accepted, although it is more restrictive than the other common interpretation that would include all the sensors. The leading biological components of biosensors are listed in Table 4.2.

TABLE 4.2 BIOLOGICAL COMPONENTS OF BIOSENSORS

Binding	Catalysis
Antibodies	Enzymes
Nucleic acid	Organelles
Receptor proteins	Tissue slices
Small molecules	Whole organisms

Currently, commercially available are biosensors for glucose, lactate, alcohol, sucrose, galactose, uric acid, alpha amylase, choline and L-lysine. All are amperometric sensors based on O_2 consumption or H_2O_2 production in conjugation with the turnover of enzyme in the presence of substrate. A urea sensor is based on urease immobilized on a pH glass electrode. In 1987, Baxter manufactured a complete glucose sensor containing disposable glucose oxidase-based electrodes, power supply, electronics and read-out in a housing, the size of a ball point pen. One places a drop of blood on the disposable electrode and a few seconds later a fairly accurate reading of blood glucose is obtained. It is widely believed that much more frequent measurement of blood glucose with correspondingly frequent adjustments of the dose of insulin delivered could significantly improve the long-term prognosis for insulin-dependent diabetes.

SENSING MODALITIES

Potential-based sensors (pH and ISE) The major impediment towards the commercialization of the potential sensors has been their long-term stability, sensor–sensor reproducibility and selectivity. Some of the first biosensors employed enzyme-catalysed reactions (such as those of pencillinase, urease and even glucose oxidase) that affect pH. By putting a pH electrode into the solution containing the enzyme it is possible to monitor that rate of enzymatic turnover. It is also possible to use pH to monitor the change in production of CO_2 by bacteria in the presence of substrates that they are capable of metabolizing. However there is always a problem for *in vivo* use of pH-based sensors related to the fact that the external environment is capable of strong buffering of pH changes, and any change in pH in the immediate environment of the sensing surface is reduced toward the bulk pH by a degree that depends on the strength of that buffering.

Electrochemical sensors Electrochemical sensors are used to determine the concentrations of various analytes in testing samples such as fluids and dissolved solid materials. Electrochemical sensors are frequently used in occupational safety, medical engineering, process measuring engineering, environmental analysis, etc. Electrochemical sensors have electrode arrays with two, three or more electrodes, which are called auxiliary electrode, reference electrode, and working electrodes. Many enzymes perform oxidation and reduction reactions and can be coupled to electrodes. The electrochemically active species in enzymes is generally a cofactor, that when bound, is not accessible to the electrode surface at which the electron transfer must take place for detection. In the case of glucose oxidase reaction, the normal biological reaction is:

$$\text{Glucose} + O_2 + H_2O \leftrightarrow \text{Gluconic acid} + H_2O_2$$

The enzyme uses an FAD coenzyme to mediate the oxidation, and the resultant $FADH_2$ is directly oxidized by O_2 to return to FAD to prepare for the next catalytic reaction. The concentration of small diffusible molecules such as O_2 is rate-limiting, so the sensor often measures not glucose but the rate at which O_2 can arrive at the enzyme to reoxidize its cofactor. There are two electrochemical ways to couple the reaction to electrodes: monitoring depletion of O_2 and protons. The latter approach is generally used to avoid direct effects of O_2 variations on the electrode, but this does not completely solve the problem. The best solution has been the use of electrochemical mediators that are at a higher concentration than O_2 and can therefore shuttle back and forth between the protein and the electrode faster than the enzyme being reduced, so that the arrival of the substrate such as glucose is always rate-limiting.

Acoustic/mechanical sensors The device is not a simple array of independent micro resonators, but these resonators are mechanically connected via a backbone to improve the energy transfer efficiency. The membrane detects and transfers the acoustic vibration to the resonator array to perform the real time frequency analysis, copying the function of the basilar membrane of the human inner ear. The advantage of this device is that the fabrication process is much simpler and less delicate. Binding of material to surfaces changes its mass, which can change either the object's resonant frequency or the velocity of vibrations propagated through it. This has allowed development of sensors called surface acoustic wave (SAW) or bulk acoustic wave (BAW) detectors that are based on oscillating crystals. Sensitive detection of analytes is relatively easy in the gas phase, and while there have been reports of selective detection of analytes using immobilized antibodies, there is still controversy as to how or if the technique works when the oscillating detectors is in contact with liquid. It is, however, unlikely that this technique will prove applicable to *in vivo* use, where some nonspecific adsorption of protein is almost unavoidable.

Thermal and phase transition sensors Based on silicon and related technologies, with an electrical output signal. The input signal can be any of the six signal types, i.e., mechanical, magnetic, chemical, radiant, thermal and electrical signals. In thermal sensors, the transduction of the input signal to the output signal is carried out in two steps. First the input signal is transduced into a thermal signal, and then the thermal signal is transduced into the electrical output signal. Chemical reactions can give up heat because they involve breaking and formation of chemical bonds, each of which has a characteristic enthalpy. There is also a strong effect of the heats of solution of the substrates and products, particularly charged species. Many enzymatic reactions release 25 to 100 kJ/mol or 5 to 25 kcal/mol. A 1-mM solution of substrate completely enzymatically converted to product with a 5 kcal/mol heat of reaction would increase in temperature by 0.005°C, which is readily measurable in laboratory conditions Table 4.3. Sensors based on this principle are in use as detectors in chromatography and in principle could be applied to almost any enzymatic reaction.

An alternative thermal approach is to use the depression in phase transition temperatures of pure compounds by dissolving in them dissimilar small molecules that prefer the fluid phase over the crystalline phase. If the temperature is known, the concentration of the small molecule can be determined by the extent of the freezing-point depression. This principle has been successfully applied to the development of a fluorescence-based fibre-optic probe for *in vivo* use.

TABLE 4.3 ENTHALPIES OF SOME ENZYMATIC REACTIONS

Enzyme	Substrate	ΔH (kJ/mol)
Catalase	H_2O_2	100
Cholesterol oxidase	Cholesterol	53
Glucose oxidase	Glucose	80
Hexokinase	Glucose	28
Lactate dehydrogenase	Sodium pyruvate	62
NADH dehydrogenase	NADH	225
Penicillinase	Penicillin-G	67
Urease	Urea	61

Biomembrane-based sensors Advantages if biomembrane based sensors approaches enabling a decrease of the detecting part dimensionality and overcoming some restrictions in the dynamics of receptor molecules on solid supported thin liquid films of lipid molecules. A complex biological system that has been applied to the development of sensors is the biological membrane and the lipids and proteins that make it up. Numerous approaches have been made to apply lipid bilayers to chemical detection, including at least to sensors for general anesthetics. Membrane receptor proteins are responsible for transducing many important biological binding events and could be used to great advantage for monitoring such chemicals as hormones, neurotransmitters and neuroactive drugs. These electrical techniques promise the most sensitive detection, as a single channel opening can be monitored electrically, but also involve some of the most difficult technical challenges, including stabilizing of the normally fragile lipid bilayer. An ancillary benefit of the use of biomembranes is that phospholipids have been reported to enhance the biocompatibility of biomaterials, so they may have a dual role in bilayer-based sensors.

Microfabrication-based sensors Microfabrication technology such as used in the microelectronics industry has been applied in the fabrication of biomedical sensors. Thin- and thick-film processing is especially well-suited to fabricating physical and chemical sensors due to the special properties of these films and the relative low costs for their production compared to other microfabrication technologies. These technologies can yield reproducible, batch-fabricated, and relatively inexpensive sensors. The use of microfabrication has become a central tool in the development of many types of sensors. The first attempt to meld semiconductors with chemical and biosensors was use of the chemical field-effect transistor, which has now become a commercial product, at least for measurement of pH. Subsequent use of microfabrication has focused on fluidic channels, optical windows, and electrodes with dimensions ranging from millimetres to micrometres. Foremost among the applications of microfabrication has been the forming of small channels in insulating materials (such as glass or plastic) for capillary electrophoresis. Although strictly speaking a chromatographic technique and not a sensor technique, integrated systems that incorporate both microcapillary electrophoresis and optical or electrical detection have so many of the functions of a sensor that the differentiation is perhaps no longer meaningful. The strong interest in making highly parallel arrays of capillaries for high-throughput screening and DNA diagnostic has prompted intense development activity in microcapillary arrays both in academics and in industry. However, there is every reason to believe that such devices will soon work their way into *in vivo* systems as well.

QUANTUM DOT TECHNOLOGY IN CANCER TREATMENT

Quantum Dots in Early Diagnosis of Cancer

Early screening of cancer is desirable as most tumours are detectable only when they reach a certain size when they contain millions of cells that may already have metastasized. Currently employed diagnostic techniques such as medical imaging, tissue biopsy and bioanalytical assay of body fluids by enzyme-linked immunosorbent assay (ELISA) are insufficiently sensitive and specific to detect most types of early-stage cancers. Moreover, these assays are labour-intensive, time-consuming, expensive and do not have multiplexing capability. On the other hand, QD-based detection is rapid, easy and economical enabling quick point-of-care screening of cancer markers. QDs have got unique properties which make them ideal for detecting tumours.

These include intense and stable fluorescence for a longer time; resistance to photobleaching, large molar extinction coefficients, and highly sensitive detection due to their ability to absorb and emit light very efficiently. Due to their large surface area-to-volume ratio, a single QD can be conjugated to various molecules, thus making QDs appealing for employment in designing more complex multifunctional nanostructures. Various types of biomarkers such as proteins, specific DNA or mRNA sequences and circulating tumour cells have been identified for cancer diagnosis from serum samples. Therefore, QD-based multiplexed approach for the simultaneous identification of many biomarkers would lead to more effective diagnosis of cancer. QDs have been covalently linked to various biomolecules such as antibodies, peptides, nucleic acids and other ligands for fluorescence probing applications. Some of the applications of QDs in biology along with their tremendous potential for *in vivo* molecular imaging have already been explored.

Advantages of Inorganic Quantum Dots over Organic Fluorophores

Compared to traditional organic fluorophores used for fluorescence labelling in biological experiments, inorganic QDs have wider applications due to their high resistance to photobleaching, which enables visualization of the biological material for a longer time. Fluorophores are highly sensitive to their local environment and can undergo photobleaching, an irreversible photo-oxidation process which makes them non-fluorescent. This is the main limitation for all studies in which the fluorophore-labelled structure has to be observed over extended periods of time. Fluorophores can be optically excited only within a narrow range of wavelengths and fluorescent emission is also restricted to a certain range of wavelengths, whereas QDs can be excited with a single light source having wavelength shorter than the wavelength of fluorescence. The fluorescence spectra of QDs are narrow, symmetric and have no red tail as observed in fluorophores. Various colours can be observed and distinguished without any spectral overlap. Therefore, multicolour labelling of different structures with QDs of different colours became possible. This multiplexed approach is of great interest in wide-ranging applications such as disease diagnosis and drug delivery. The field of QDs is multidisciplinary one as personnel from different scientific disciplines, i.e., chemistry, physics, biology and medicine are working together to harness their potential. Their employment for the detection and treatment of cancer is one such application which is of paramount importance.

QUANTUM DOT TECHNOLOGY

QDs are inorganic semiconductor nanocrystals having typical diameter between 2–8 nm that possess unique luminescent properties. They are generally composed of atoms from groups II and VI elements (e.g. CdSe and CdTe) or groups III and

V elements (e.g. InP and InAs) of the periodic table. Their physical dimensions are smaller than the exciton Bohr radius that leads to quantum confinement effect, which is responsible for their unique optical and electronic properties.

Synthesis of Quantum Dots

High-quality QDs have been synthesized by various approaches. But usually their synthesis is carried out in organic solvents such as toluene or chloroform at higher temperatures in the presence of surfactants. But the surfactant-coated particles are not soluble in water as they have polar surfactant head group attached to the inorganic core of QD and the hydrophobic chain protruding into the organic solvent. Usually, all experiments with cells involve water-soluble materials. Therefore, various strategies have been developed to make them water-soluble, where either the surfactant layer is replaced or coated with additional layer such as hydrophilic or amphipathic polymers. The hydrophobic coating of surfactant is replaced by ligand molecules carrying functional groups at the one end that bind to the QD surface, and hydrophilic groups at the other end that make the QDs water-soluble. The employment of amphiphilic polymers as an additional coating on QD surface has also been reported. The hydrophobic tail of the polymer reacts with the hydrophobic surfactant layer on QD surface whereas the hydrophilic groups of the polymer on the outer end impart water solubility. QDs have also been encapsulated in phospholipid micelles to make them water-soluble.

Properties and Applications of Quantum Dots

The most commonly used QD system is the inner semiconductor core of CdSe coated with the outer shell of ZnS. The ZnS shell is responsible for the chemical and optical stability of the CdSe core. QDs can be made to emit fluorescent light in the ultraviolet to infrared spectrum just by varying their size. The wavelength of fluorescence of the QD depends on its energy gap (i.e., the difference between the excited and the ground state) which is determined by the size of the QD. QDs have narrow spectral line widths, very high levels of brightness, large absorption coefficients across a wide spectral range, high photostability and capability of multiplexed detection. They are very bright and stable even under complex *in vivo* conditions that make them suitable for advanced molecular and cellular imaging, drug delivery and for highly sensitive bioassays and diagnostics. Highly sensitive real-time imaging with greater resolution and tracking of single receptor molecules on the surface of living cells have been made possible by

QD bioconjugates. Various applications of quantum dots are stated in Figure 4.7. In most of the cases, functional QD conjugates for cancer detection are composed of a semiconductor core (CdSe, CdTe); an additional shell such as ZnS in the case of CdSe QDs having a higher band gap than CdSe to increase quantum yield; a water-soluble hydrophilic coating; and, functionalized antibodies or other biomolecules complementary to the target cancer markers at the tumour sites.

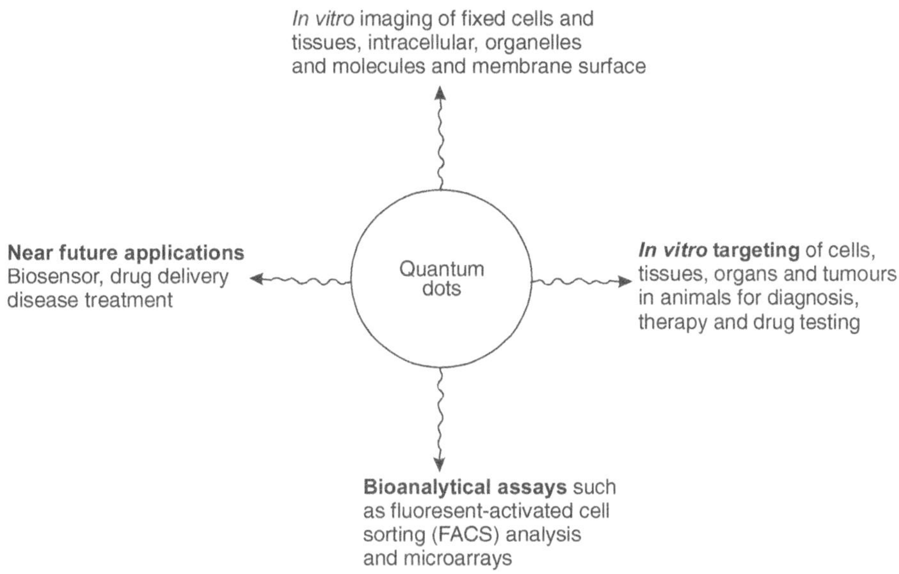

FIGURE 4.7 VARIOUS APPLICATIONS OF QUANTUM DOTS

OVERCOMING THE TOXIC NATURE OF QUANTUM DOTS

The native QDs made up of semiconductor nanoparticles are toxic in nature. It has been observed that CdSe QDs are highly toxic to cells exposed to UV for a longer time as UV dissolves the CdSe, thereby releasing toxic cadmium ions. However, polymer-coated QDs are non-toxic in the absence of UV as demonstrated by *in vivo* studies. It has also been shown that the micelle-encapsulated QDs injected into the frog embryo did not affect its development. Therefore, QDs are normally encapsulated inside the outer coating of amphiphilic polymers to make them water-soluble and resistant to chemical or enzymatic degradation. They are typically synthesized in organic solvents such as tri-*n*-octyl-phosphine oxide (TOPO) and hexadecylamine, having long alkyl chains and high boiling points, to prevent the formation of aggregates. In the recent years, there has been a great development to modify the surface chemistry of QDs to make them water-soluble. Most commonly,

QDs are linked to polyethylene glycol (PEG) or similar ligands to make them biocompatible and to reduce nonspecific binding. They are made specific to the target site by conjugating them to various bioaffinity ligands such as peptides, antibodies, oligonucleotides, etc. using different strategies. A possible schematic representation of the QD bioconjugate for the detection of tumor cell biomarkers is shown in Figure 4.8.

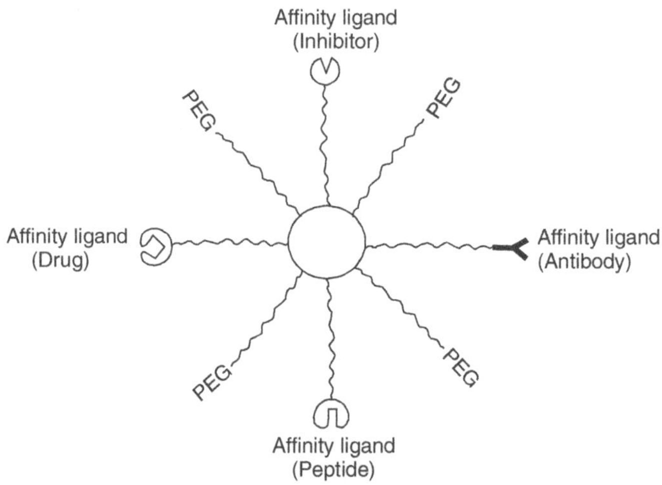

FIGURE 4.8 MULTIFUNCTIONAL QDS ARE CONJUGATED TO VARIOUS AFFINITY LIGANDS SPECIFIC FOR THE CELL BIOMARKER

BLINKING BEHAVIOUR OF QUANTUM DOTS

Nirmal *et al.* discovered for the first time that QDs shows a blinking behaviour, i.e., intermittent on–off emission upon continuous excitation, which was attributed to Auger ionization. The principle of this behaviour is not well understood even today. But it is a concern only when a signal from individual QD is required during the analysis such as flow cytometry applications. In such cases, it may be possible that the emission from the individual QD might be off due to "blinking" thus leading to the missing of signal at the detector. But generally in most of the applications such as in cell-based assays, there are more than one QD involved and even if some QDs are blinking, others are giving signal for the final detection and thus, no signal will be missed by the detector. One way of counteracting the reduced quantum yield due to blinking is to grow a shell of a few atomic layers of a material with a larger band gap on top of the QD core.

Effect of Surface Functionalization on the Optical Properties of Quantum Dots

Fundamental studies have revealed that luminescence of QD is very much sensitive to the surface functionalization procedures as the interactions of the molecule with the QD's surface change the surface charges on the QD. But many of the QD-based probing applications are based on the change in fluorescence of QD after the interaction of the target analyte molecules with the biomolecules functionalized on the QD surface. It has been well reported that the surface functionalization of QDs improves their solubility. But it could reduce their quantum efficiency as well. This has been demonstrated in the case of mercaptoacetic acid-treated QDs where the quantum efficiency was reduced drastically. But protein-functionalized quantum dots tend to retain their quantum efficiency and offer longer shelf life. They can also be further functionalized with multiple functional groups without decreasing their quantum efficiency.

Measurement System for Observing and Tracking Quantum dots

Single QDs can be observed and tracked for greater time duration up to a few hours with confocal microscopy, total internal reflection microscopy or epifluorescence microscopy. The scheme of the fluorescent imaging employing QDs as labels and its measurement has been described by Gao *et al.* They employed a whole-body macro-illumination system with wavelength-resolved spectral imaging, which allows high sensitivity detection of molecular targets *in vivo* and also employed the wavelength-resolved spectral imaging system having software that separated autofluorescence from quantum dot signals.

Advances in Quantum Dot Technology for the Diagnosis of Cancer

In the early stages, QDs were employed for several imaging applications in place of organic dyes. But the tremendous potential of these materials was realized when it was observed that they kept on emitting intense fluorescent light for weeks. This was a major technological advancement for microscopic imaging, which helped in unfolding many cellular processes. In the subsequent stages of development, researchers developed a keen interest in the QD technology and started exploring their applications in different fields. Different QDs composed of the same material but of different sizes had been made, which can generate different colours after activation by light of a single wavelength. It was then demonstrated that QDs tagged with biomolecules such as antibodies, peptides, etc. can be employed to detect specific molecules on the cell surface or inside the cell.

Researchers have started the exploration of QDs just from the last two decades. The field is still in its infancy but it has captivated scientists and engineers due to the unique optical and electronic properties of QDs. QDs have revolutionized the field of molecular imaging. The forthcoming years would see their potential applications in different fields. One of the major areas of impact is surely the intracellular imaging of live cells. The technology will provide new insights in understanding the pathophysiology of cancer, and in imaging and screening tumours. QDs will definitely be one of the components of the envisioned multifunctional nanodevices that can detect diseased tissue, provide treatment and report progress in real time.

NANOTECHNOLOGY: A FOCUS ON ADVANCED DRUG DELIVERY SYSTEM

An important and long-term goal of the pharmaceutical industry is to develop therapeutic agents that can be selectively delivered to specific areas in the body to maximize the therapeutic index. Drugs, given systemically, provide a profound beneficial effect but can also exhibit adverse reactions. Historically, cancer chemotherapy agents have been well-known examples of achieving balance between efficacy and toxicity. Cytotoxic compounds can be highly effective in destroying cancer cells but may also damage normal cells resulting in possible adverse and potentially life-threatening effects. The "magic bullet" concept, first theorized by Paul Ehrlich in 1891, represents the first early description of the drug-targeting paradigm. The aim of drug targeting is to deliver drugs to the right place, at the right concentration, for the right period of time. As drug characteristics differ substantially in chemical composition, molecular size, hydrophilicity, and protein binding, the essential characteristics that identify efficacy are highly complex. All of these are investigated to bring a new compound to market although only a fraction reaches active clinical use.

Many promising new compounds are compromised by poor physiochemical properties that leads to poor solubility and biodistribution and therefore the drug does not interact with the site of action. Poor oral absorption (i.e., proteins and peptides), low solubility at physiological pH, insufficient cellular uptake, and rapid drug elimination are impediments for drug development. New drug candidates must provide evidence that they reach the site of action and have an effect. The field of drug delivery designs carriers, excipients, and solubilizers to transport drugs to the site of action. An answer to poor drug physiochemical properties is to associate the drug with a pharmaceutical carrier [i.e., a drug delivery system (DDS)]. A DDS can enhance a drug pharmacokinetics and cellular penetration. Moreover, obstacles arising from low drug solubility, degradation, fast clearance rates, nonspecific toxicity, and inability to cross biological barriers may also be addressed by a DDS. To be useful, a DDS is required to be biocompatible with processes in the body as well as with the drug to be delivered. Overall, the challenge of increasing a drug's therapeutic effect, with a concurrent minimization of side effects, can be optimized through proper design and DDS engineering.

Nanoparticles as Drug Carriers

Nanoparticles were first developed approximately 35 years ago. They were initially developed as carriers for vaccines and cancer chemotherapy agents. Nanoparticles are stable, solid colloidal particles consisting of biodegradable polymer or lipid materials and range in size from 10 to 1,000 nm. Drugs can be absorbed onto the particle surface, entrapped inside the polymer/lipid, or dissolved within the particle matrix. An example of a DDS is a liposome. Liposomes are closed bilayered phospholipids first designed 40 years ago. Table 4.4 lists examples of FDA-approved nanoparticle DDS. These were collectively designed to be taken up and delivered by mononuclear phagocytes (MP).

The majority of pharmaceutical research into the utility of nanoparticle DDS has been in the area of oncology. Nanocarriers can concentrate preferentially in tumour masses, inflammatory sites, and infectious sites by virtue of the enhanced permeability and retention (EPR) effect on the vasculature. Characteristics of the EPR effect include extensive angiogenesis, defective vascular architecture, impaired lymphatic drainage/recovery system, and increased production of a number of permeability mediators. These EPR characteristics are essential components of solid tumours and are not associated with normal tissues or organs. The EPR effect provides an opportunity for more selective targeting of nanoparticle (lipid or polymer conjugated anticancer drugs) to the tumour. Additionally, it is possible to fabricate nanoparticles with several different drugs and allow for the nanocarriers to selectively deliver the drugs to the malignant tissue, be ingested by the malignant cells, and then release the now intracellular anticancer drugs killing malignant cells. Nanocarriers can also use more than one drug within the particles—an anticancer drug incorporated within a biodegradable colloidal shell and another anti-angiogenesis drug that is placed within a lipid layer that surrounds the colloidal shell.

TABLE 4.4 FDA-APPROVED NANOPARTICLES FOR DRUG DELIVERY

Drug or chemotherapeutic agent	Indication
Liposomal Amphotericin B	Fungal infection and leishmaniasis
PEG-adenosine deaminase	Severe combined immunodeficiency disease
PEG stabilized liposomal doxorubicin	Kafocy sarcoma and refractory ovarian cancer
Liposomal cytosine arabinoside	Lymphomatous meningitis and neoplastic meningitis
Interleukin 2-diphtheria toxin fusion protein	Cutaneous T-cell lymphoma
Liposomal vertiporfin	Wet macular degeneration
PEG interferon α-2b	Hepatitis C
PEG-granulocyte colony stimulating factor	Chemotherapy-associated neutropenia
Protein bound paclitaxel	Metastatic breast cancer
PEG L-asparagine	Acute lymphocytic leukaemia
PEG aptanib	Hepatitis C
Penetrexed	Malignant pleural mesothelioma

When the nanocarrier is administered intravenously, the malignant cells internalize the nanoparticle. Once intracellular, the anti-angiogenesis drug within the lipid layer of the nanocarrier is released, inhibiting the mediators for blood vessel production. Then, the core anticancer drug within the colloidal shell is released, effectively killing the malignant cell. All of this can be fabricated in a "nanocell" that is less than 250 nm. This is truly an optimized delivery of anticancer drugs to the tumour site. Doxorubicin and paclitaxel are FDA-approved anticancer drugs that are limited by physiochemical properties that make them difficult to administer or have intolerable side effects. Paclitaxel is a difficult drug to keep in an aqueous solution. Cremophor and alcohol are added to the product as solubilizers to prevent paclitaxel from precipitating out of aqueous solution. However, toxicities (i.e., hypersensitivity reaction and neurotoxicity) associated with Cremophor are problematic for patients to tolerate.

Additionally polyethylene glycol (PEG)-coupled paclitaxel has been responsible for some infusion reactions. Therefore, albumin-bound paclitaxel has been developed to enhance product solubility and improve patient tolerability.

This formulation has an improved side effect profile compared to the conventional paclitaxel. Doxorubicin has a potential life-threatening side effect, cardiotoxicity, related to the cumulative dose. When administered to patients, pegylated liposomal doxorubicin dose and frequency of administration are reduced because of the improved penetration to the malignant tissue. The cumulative dose is reduced. Therefore, the potential life-threatening cardiotoxicity associated with this drug is reduced. Using pegylated doxorubicin in patients with breast metastases has shown improved tissue concentrations and survival at a median initial dose of 42.5 mg/m^2 every 4 weeks. The development of the liposome-based, slow-release formulation of cytarabine (ara-C) has been a step forward in the treatment of lymphomatous meningitis. The product is encapsulated in multivesicular, lipid-based particles with diameters of 3–30 μm and consists of numerous nonconcentric vesicles. Treatment modalities for another tumour, malignant pleural mesothelioma, have not produced any impact on median survival, and many toxicities associated with standard chemotherapy have been found to be harsh and result in increased mortality and morbidity.

Liposomal penetrexed was recently approved for mesothelioma. Other liposomal anticancer drugs are currently in various stages of development. Polyethylene glycol-filgrastim, used to treat neutropenia (low WBC) associated with cancer chemotherapy, is another example of nanoparticulate DDS. As a result of the addition of PEG, circulating proteins do not complex with PEG filgrastim and eliminate it from the circulation. This keeps PEG-filgrastim in the circulation longer than filgrastim without pegylation. Therefore, this form of filgrastim can be given less often compared to conventional filgrastim. A clinical study was performed to evaluate the efficacy of PEG-filgrastim compared to conventional filgrastim in the development of febrile neutropenia in patients with breast cancer. In the PEG-filgrastim group, significantly fewer patients (1%) developed neutropenia compared to conventional filgrastim (17%).

The ERP effect can be exploited for infectious diseases or other inflammatory diseases. Pegylated interferon alpha 2a is used in the treatment of patients with hepatitis C. The recommended dose, a once-a-weekly injection, is primarily attributable to the PEG masking the protein from MPS clearance, thereby allowing serum concentrations to be relatively constant over the dosing period. An example of an experimental DDS is the drug rolipram, an anti-inflammatory drug, which was incorporated into nanoparticles and tested in a rat model of inflammatory bowel disease. Animals that received the nanoparticle formulation demonstrated a better response based on myeloperoxide activity and lack of weight loss compared to rolipram solution. Incorporating aminoglycoside antibiotics into nanoparticles or microparticles have reduced drug-induced nephrotoxicity.

Amikacin-liposome product has undergone clinical trials in an attempt to maximize the therapeutic window and minimize the toxicity of this antibiotic. Overall therapeutic index is improved compared to conventional amikacin. Amikacin encapsulated in unilamellar liposomes (MiKasome®) has prolonged circulation time and sustained efficacy in animals infected with *Mycobacterium* spp., *Klebsiella* spp., and *Pseudomonas* spp. It is hypothesized that the acidic intracellular pH promotes release of the amikacin after phagocytosis by the macrophage effectively delivering drug to the intracellular pathogen. Indeed, treatment of intracellular organisms such as mycobacterial organisms using liposomal technology and the macrophage as the carrier is improved compared to free drug.

Important examples of effective DDS used outside of the treatment of oncology and infectious diseases include advances in the treatment of eye disease. Age-related macular degeneration causes progressive loss of central vision leading to blindness that is a result of abnormal growth of blood vessels under the retina and macula. The walls of these blood vessels easily rupture and bleed into the surrounding retinal tissues causing the macula to bulge and distort/destroy central vision. Verteporfin therapy is part of a two-step process that incorporates intravenous liposomal verteporfin along with light from a diode laser that is FDA-approved for wet macular degeneration. Verteporfin, once activated by light, is highly reactive and produces singlet oxygen and reactive oxygen radicals that disrupt the abnormal neovascular endothelium and cause occlusion of the leaky vessels. The damaged endothelium releases vasoactive factors and procoagulants, resulting in platelet aggregation, fibrin clot formation, and vasoconstriction. Another drug, pegaptanib sodium, has recently gained FDA approval for the same indication. This drug is a pegylated synthetic oligonucleotide that acts as an antagonist of vascular endothelial growth factor isoform. This isoform is associated with ocular disease. Small molecular weight drugs fabricated into a DDS using nanoparticles have been shown to be useful for the treatment of malignancies and infectious diseases. Other investigators have fabricated anti-HIV drug nanoparticles.

Evading Phagocytosis

Colloids larger than a few micrometres accumulate in the lung capillaries and remain there. Sufficiently smaller structures can passively migrate through intercellular junctions of healthy endothelium or may be removed by the bone marrow. Circumventing the MPS or mononuclear phagocytic system (monocytes, macrophages, microglia, Kuppfer liver cells, and spleen red pulp macrophages) is an additional step toward successful targeted delivery. The MPS migrate throughout the body guided by chemoattractant gradients to encounter and ingest foreign particles or organisms. After encountering foreign particles, the MPS recognize specific surface ligands and become activated. Activation occurs through intracellular cytokine priming mechanisms. Activated cells then ingest the particles and partition them within intracellular phagosomes that contain granule-associated lysosomal enzymes and reactive oxide intermediates that all function to degrade the ingested material. The activated macrophages secrete cytokines that recruit other macrophages and neutrophils into the area. Hepatic and splenic cells can rapidly phagocytize polymeric nanoparticles, and fabrication of nanoparticles that can avoid or minimize phagocytosis is important for sustained circulation and effect.

Drug carriers, recognized as foreign objects, are removed from the circulation by surface interactions between cells and the carrier. Blood protein absorption onto colloids is dependent on surface physicochemical properties (size, shape, charge density, hydrophobicity). The methods used (spray-dried or w/o/w emulsion solvent evaporation) to create the particles may influence protein absorption. Opsonins are proteins that promote the activation of the complement system and assist in phagocytic uptake by macrophages. Prevention of opsonin adsorption via carrier surface modification has proven to be a successful strategy for sustained circulation. Among the surface modifiers examined, polyethylene glycol (PEG) is certainly the most well known. PEG chains form a protein-repellent layer around the carrier. It is the limited concentration of blood opsonin that partly contributes to the prolonged circulation time for PEG particles. In contrast to PEG, dysopsonins are proteins that suppress recognition by macrophages.

Favourable interactions between surface-bound polymers and blood dysopsonins can also contribute to the prolonged circulation time. Optimization of the use of dysopsonins to diminish blood opsonin absorption would be advantageous at increasing the residence time of nanoparticles. This is a promising approach toward higher plasma retention. Generally active targeting is accomplished by attachment of specific molecules on the carrier's surface, thereby enhancing the binding and interactions with antigens or receptors expressed on target cell populations. One approach to deliver compounds to specific target sites is to attach ligands to the carrier. Some ligands include monoclonal antibodies or peptides. The choice of the appropriate ligand is based on its specificity, stability, availability, and selectivity on the target cells. Liposomes have been extensively studied for this purpose and can deliver a high drug payload, protect the drug from degradation, and improve pharmacokinetics.

With the exception of topical administration and local (joint) injection of drugs, most other routes are, by nature, systemic routes of administration. Currently, a typical dose of a drug is usually tens to hundreds of micrograms. For example, the adult EpiPen used for anaphylactic shock delivers 300 μg of epinephrine. Fentanyl, a widely used narcotic, is usually given as 50 μg for a typical 70-kg adult. Common antibiotics are considerably higher for adults. A device that contains a reservoir that is a cube of 1 mm contains a volume of 1 μl, and a cube 100 μm on a side holds a volume of 1 nl. Small device fabrication would allow for the implantation of a reservoir of localized drug that could be released over time. A good example of this is the clinical use of a device that releases leuprolide acetate over a 12-month time frame that is 45 × 4 mm and holds 74 mg of drug. Using nanotechnology, a drug delivery platform has been fabricated using poly(vinyl alcohol) microgels that entrap dexamethasone-loaded poly-(DL-lactide-co-glycolide) (PLGA) microspheres for controlled delivery over a 1-month period. The rate-limiting step for drug delivery devices that release drugs over an extended period is the associated microelectromechanical system (MEMS) themselves.

Using radio frequency, magnetic field, or ultrasound for activating drugs are interesting options to control the delivery of a drug. Ferromagnetic rods could be placed within a target tumour and thermal treatment be initiated by using radio frequency or ultrasound. In an experiment, magnetic liposomes containing doxorubicin were administered to osteosarcoma-bearing hamsters. When the tumour limb was placed between two poles of a 0.4-T magnet for 60 min, it resulted in a fourfold increase in tumour drug levels. Ferromagnetic microparticles are also used as magnetic resonance imaging contrast agents. Liposomal contrast agents have been utilized for experimental diagnostic imaging of liver, spleen, brain, cardiovascular system, tumours, inflammation, and infection. Using polymers that contain metals (^{111}In or Gd) enhances the signal intensity. For magnetic resonance imaging (MRI), iodinated organic compounds can be incorporated into the liposomal membrane. A preliminary investigation with infectious organisms has shown systemic locations where nanoparticles migrate. Intramuscular *Staphylococcus aureus* abscess was induced in rats and colloidal gold nanoparticles have been shown to concentrate in the blood vessels surrounding the abscess, as well as in the liver and spleen. Another nanoparticle used in cell trafficking studies and tumour targeting are PEG-superparamagnetic iron oxide and PEG-quantum dots.

Intracellular Trafficking

Once the drug delivery vehicle has reached the site, subsequent drug release may occur within the extracellular space, or following internalization of the carrier into the target cell. Drugs with intracellular action, incapable of crossing cell membranes, need to be assisted in reaching their target. Cellular uptake mechanisms vary according to cell type, physicochemical properties of the internalized compound, and mode of activation. Intracellular targeting is feasible through the use of ligands that trigger receptor-mediated endocytosis. Studies on PLGA nanoparticles and localized delivery of plasmid DNA, proteins and peptides and small molecular weight drugs, revealed that PLGA nanoparticles uptake is both concentration- and time-dependent. The uptake of nanoparticles occurs by nonspecific fluid phase endocytosis and can be saturated as uptake decreases with higher doses. PLGA nanoparticles are transported into primary endosomes and sorted to either recycling endosomes or secondary endosomes. In the acidic environment of the secondary endosome, the nanoparticle surface changes from anionic to cationic leading to the escape of the nanoparticles into the cytoplasm. While intracellular nanoparticle levels fall, extracellular nanoparticle levels may not fall rapidly. A previous study has demonstrated that when nanoparticles are delivered locally, small molecular weight drug levels in the associated tissues are sustained for up to 7–14 days.

Inhaled DDS

There are difficulties associated with the treatment of tuberculosis (TB). The organism grows slowly and therefore, anti-TB drugs must be given for an extended period of time to be effective. Because of this extended treatment, adherence is problematic for patients. More than 80% of TB cases involve the pulmonary tissue alone. High doses of anti-TB drugs are required because only a small fraction of the total dose reaches the lungs after oral administration. Adherence to treatment and patient outcome could be optimized with the introduction of long-acting drug formulations that would release drugs at a slow and consistent rate. Investigators have utilized drug-loaded nanoparticles and investigated an inhaled delivery system in a guinea pig model. The particles ranged from 186 to 290 nm in diameter, and the anti-TB drugs that were loaded included rifampin, isoniazid, and pyrazinamide. A single nebulization of this formulation was able to maintain a therapeutic plasma concentration for 6–8 days and a lung concentration for 9–11 days. Advantages of this form of therapy, if the nanoparticles can be delivered as a nebulization to humans, are readily apparent. Other avenues of research to optimize the delivery of nanoparticles to the lung tissue are being actively investigated.

In vivo Imaging and Diagnostics

The utility of quantum dots for *in vivo* imaging, cellular functional assays, and diagnostics has the potential to revolutionize medical research. Quantum dots are nanocrystals or nanotubes and are only a few nanometres in diameter and made of semiconductor materials. Quantum dots work by fluorescence resonance (blinking) visible at the single-molecule level. These dots can be observed and tracked over an extended period of time with confocal microscopy, total internal reflection microscopy, or wide-field epifluorescence microscopy. A utility of quantum dots is their ability to tag these to a target molecule (i.e., ligand, DNA, antibody, etc.) used in cellular mechanisms, for *in vivo* imaging of a tumour, or to have a read-out of a functional assay for diagnostic purposes. Many other uses are currently being investigated for cellular assays, imaging, and diagnostics.

The Future of Nanotechnology Drug Delivery

The therapeutic advantages of nanotechnology-derived drug delivery are becoming apparent and will soon be associated with every route of drug administration. The advantages over current treatment modalities include lower drug toxicities, improved bioavailability, reduced economic costs of treatment, and increased patient adherence to treatment. The medical management of malignancies has already been greatly impacted by nanotechnology, but soon other medical specialties will utilize these novel forms of drug delivery to achieve optimal treatment success.

Additionally, innovative research and development of more therapeutically effective nanocarriers will continue including improved forms of polymer–drug conjugates, liposomes, dendrimers, micelles, polymeric vesicles, and nanocapsules. Finally, implantable drug delivery systems will open up many more opportunities for nanotechnology utilization. Optimized drug release from implantable delivery systems is preferable to intravenous administration. The extended duration of action, reduced frequency of redosing, and improved patient acceptance are all positive attributes of implantable drug delivery systems. Although the future of nanotechnology is promising, it is important to consider the toxicological aspects of nanoparticles. Toxicity screening of nanotechnology products should include physiochemical characterization of the nanomaterial, *in vitro* assays, and *in vivo* studies to assure the public that these materials will not influence cellular reactions within the human body.

Nanotechnology will become more commonplace in the delivery of drug therapy as pharmaceutical innovation continues. The increasing interest in this field of drug research demonstrates the implicit promise and potential of nanotechnology-derived drug delivery paradigms. Experimental findings provide strong support for the advancement of nanotechnology-derived drug delivery for antiretroviral drugs. The use of nanotechnology for the purposes of diagnostics and imaging will also become increasingly important.

The use of MEMS to mimic secretion of hormones, insulin, and other closed systems is a central goal of nanotechnology-derived devices. Additionally, the release of narcotic drugs for severe pain relief would be advantageous if an implantable device can be fabricated. Research into the fabrication of these much needed devices may soon lead to possible alternatives to current treatment modalities. The future of quantum dots for imaging, diagnostics, and cellular applications will continue to develop. The quantum dots possibly could be injected intravenously and target a diseased tissue (cancer). These nanosized particles could be used as contrast agents for MRI, positron emission tomography (PET), and computed tomography to view the malignant tissue. An optical biopsy could be performed that confirms the pathology. A nanoparticle DDS could then be utilized to treat the malignant cells with minimal side effects to the patient.

Certainly, there will be other uses associated with this evolving technology. However, in order for nanotechnology to continue the evolution of medical management, it will require the multidisciplinary approach from both basic and clinical research backgrounds to achieve sustained innovation. Many clinical investigators will collaborate with material scientists, engineers, and polymer chemists to bring this technology to the forefront in the quest to optimize essential medical therapies for patients. It is quite likely that nanotechnology will become the next frontier of medical research.

DNA-BASED ARTIFICIAL NANOSTRUCTURES AND THEIR APPLICATIONS

The integration of nanotechnology with biology and bioengineering is producing many advances. The essence of nanotechnology is to produce and manipulate well-defined structures on the nanometre scale with high accuracy. Conventional technologies based on the "top-down" approaches, such as the photolithographic method, are difficult to continue to scale down due to real physical limitations including size of atoms, wavelengths of radiation used for lithography, and interconnect schemes. While engineers and scientists have long aspired to controllably manipulate structures at the micrometre and nanometre scale, nature elegantly performs these tasks and assembles structures with great accuracy and high efficiency using specific biological molecules. Biological molecules, such as DNA, have shown great potential in fabrication and construction of nanostructures and devices. DNA molecules can be used for the assembly of devices and computational elements, for the assembly of interconnects, or as the device element itself. There are several advantages to use DNA for these constructions.

First, DNA is the molecule whose intermolecular interactions are the most readily programmed and reliably predicted: Docking experiments reduce to the simple rules that A pairs with T, and G pairs with C. Thus, the very properties that make DNA so effective as genetic material also make it an excellent molecule for programmed self-assembly. Second, DNA of arbitrary sequences is available by convenient solid support synthesis. The needs of the biotechnology industry have also led to reliable chemistries to produce modifications, such as biotin groups, fluorescent labels, and linking functions. Third, DNA can be manipulated and modified by a large battery of enzymes that include DNA ligases, restriction endonucleases, kinases, and exonucleases. On the other hand, nanotechnology has helped the development of novel biosensors for biological and medical applications. Nanobioconjugates that consist of various functional nanoparticles linked to biological molecules have been used in many areas such as diagnostics, therapeutics, sensors, and bioengineering. Detection methods based on these nanobioconjugates show increased selectivity and sensitivity as compared with many conventional assays that rely on molecular probes.

Fundamentals of DNA

DNA is the basic building block of life. Hereditary information is encoded in the chemical language of DNA and reproduced in all cells of living organisms. The double-stranded helical structure of DNA is key to its use in self-assembly applications. Each strand of the DNA is about 2 nm wide and composed of a linear chain of four possible bases (adenine, cytosine, guanine, and thymine) on a backbone of alternating sugar molecules and phosphate ions. Each unit of a phosphate, a sugar molecule, and base is called a nucleotide and is about 0.34 nm long. The specific binding through hydrogen bonds between adenine (A) and thymine (T), and cytosine (C) and guanine (G) can result in the joining of two complementary single-stranded DNA to form a double-stranded DNA. There are two hydrogen bonds between A-T pairs and three hydrogen bonds between G-C pairs. The phosphate ion carries a negative charge in the DNA molecule, which results in electrostatic repulsion of the two strands. In order to keep the two strands together, positive ions must be present in the solution to keep the negative charges neutralized.

The joining of two complementary single strands of DNA through hydrogen-bonding to form a double-stranded DNA is called hybridization. If a double-stranded DNA is heated above a certain temperature, the two strands will start to dehybridize and eventually separate into single strands. The centre temperature of this transition is called the melting temperature, T_m, which is a sensitive function of environmental conditions such as ionic strength, pH, and solvent conditions. As the temperature is reduced, the two strands will eventually come together by diffusion and rehybridize to form the double-stranded structure. These properties of the DNA can be utilized in the ordering and assembly of artificial structures if these structures can be attached to DNA.

Attachment of DNA to Surfaces

The first step toward DNA-based nanotechnology is to attach DNA molecules to surfaces. So far, the most widely used attachment scheme utilizes the covalent bond between sulphur and gold. The bonding of the sulphur head group to the gold substrate is in the form of a metal thiolate, which is a very strong bond

(~44 kcal/mol), and hence the resulting films are quite stable and very suitable for surface attachment of functional groups. For example, the DNA molecule can be functionalized with a thiol (S—H) or a disulphide (S—S) group at the 3′ or 5′ end. It should be noted that there are also other strategies to attach DNA to surfaces, for example, the covalent binding of DNA oligonucleotides to a preactivated particle surface and adsorption of biotinylated oligonucleotides on a particle surface coated with avidin. These attachment schemes have served as the fundamental base for DNA-related self-assembly of artificial nanostructures.

DNA-BASED NANOMATERIALS AS BIOSENSORS

In recent years, there have been significant interests in using novel solid-state nanomaterials for biological and medical applications. The unique physical properties of nanoscale solids (dots or wires) in conjunction with the remarkable recognition capabilities of biomolecules could lead to miniature biological electronics and optical devices including probes and sensors. Such devices may exhibit advantages over existing technology not only in size but also in performance. Sequence-specific DNA detection is an important topic because of its application in the diagnosis of pathogenic and genetic diseases. Many detection techniques have been developed that rely upon target hybridization with radioactive, fluorescent, chemiluminescent, or other types of labelled probes. Moreover, there are indirect detection methods that rely on enzymes to generate colorimetric, fluorescent, or chemiluminescent signals. Introduction of a single-stranded target oligonucleotide (30 bases) into a solution containing the appropriate probes resulted in the formation of a polymeric network of nanoparticles with a concomitant red-to-pinkish/purple colour change. Hybridization was facilitated by annealing and melting of the solutions, and the denaturation of these hybrid materials showed transition temperatures over a narrow range that allowed differentiation of a variety of imperfect targets. Transfer of the hybridization mixture to a reverse-phase silica plate resulted in a blue colour upon drying that could be detected visually. The unoptimized system can detect about 10 femtomoles of an oligonucleotide. This method has many desirable features including rapid detection, a colorimetric response, good selectivity, and a little or no required instrumentation.

Maxwell *et al.* also reported that biomolecules and nanoparticles can be both structurally and functionally linked to create a new class of nanobiosensors that is able to recognize and detect specific DNA sequences and single-base mutations in a homogeneous format. The principle of this detection method is illuminated in Figure 4.9. At the core of this biosensor is a 2.5 nm gold nanoparticle that functions as both a nano-scaffold and a nano-quencher. Oligonucleotide molecules labelled with a thiol group are attached at one end of the core and a fluorophore at the other end. This hybrid construct is found to spontaneously assemble into a constrained arch-like conformation on the particle surface. In the assembled state, the fluorophore is quenched by the nanoparticle. Upon target binding, the constrained conformation opens and the fluorophore leaves the surface because of the structural rigidity of the hybridized DNA (double-stranded), and fluorescence is restored. This structural change generates a fluorescence signal that is highly sensitive and specific to the target DNA. A major challenge in the area of DNA detection is the development of methods that do not rely on polymerase chain reaction or comparable target-amplification systems that require additional instrumentation and reagents.

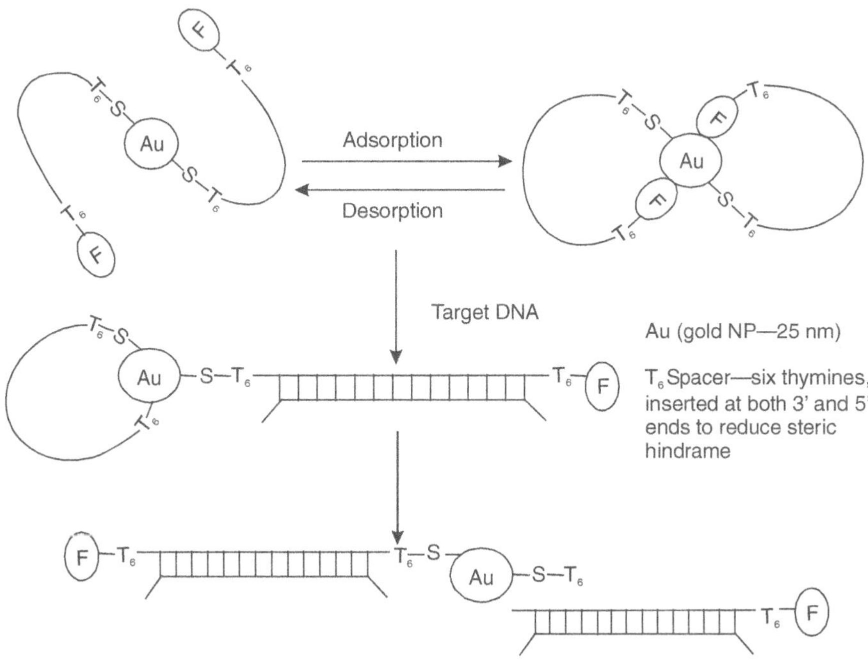

FIGURE 4.9 OPERATING PRINCIPLES OF NANOPARTICLE BASED PROBES

Due to its unique recognition capabilities, physicochemical stability, mechanical rigidity, and high-precision processibility, DNA is a promising material for "biomolecular nanotechnology." The study of DNA-based nanostructures is hence an attractive topic and describes the utilization of DNA for preparing nanostructured materials and use of such nanostructures for biological and medical applications.

Though significant progress has been made, the study of DNA-based nanostructures is still at its early stage. The catalytic, electrical, magnetic, and electrochemical properties of such structures have not yet been systematically investigated, and they, therefore, represent new frontiers in this field. It is anticipated that new phenomena and useful structures will continue to emerge over the next few years. Advanced study in this field will not only provide valuable fundamental information about the collective physical and chemical properties of nanoparticles and DNA, but also may provide access to new and useful electronic and photonic materials applicable to the industry.

DIAGNOSTIC IMAGING TECHNIQUES

It is not enough that we build instruments to produce magnified and resolved images of the microscopic world. We must able to detect these images, to visualize the information, record, and archive images, and make useful quantitative measurements of our observations. This is the science of imaging.

Digital Imaging Devices

Solid-state imaging devices are now a common part of our everyday world. Digital imaging devices are found in photocopiers, scanners, professional and consumer-grade video cameras, surveillance cameras, etc. The most common digital imaging device is the charge-coupled device (CCD). A CCD is a silicon chip that comprises a physical array of photodetectors, which correspond to the picture elements (pixels) in the image. The performance of a scientific imaging device depends on several important parameters. The resolution is determined by the number of sensing elements as well as the physical size of each photodiode. The sensitivity is determined by signal/noise ratio.

Molecular imaging

The noninvasive visualization of specific molecular targets, pathways and physiological effects *in vivo* would facilitate the study about the cell function as well as extracellular synthesis and remodelling in the assessment of biomaterials–tissue interactions and evolution of engineered tissues. Macroscopic optical molecular imaging techniques using specific molecular biomarker targets deep inside living animals has become possible as a result of a number of advances, including biocompatible fluorescent probes and mathematical modelling of photon propagation in tissue. For example, the proteolytic activity in atherosclerotic plaque that potentially causes its rupture, resulting in myocardial infarction, has been visualized, raising the potential of catheter-based monitoring. Using this technology, cell structure and function can be visualized in high spatial and temporal resolution, thereby permitting unique visualization of molecular processes *in vivo*.

Nanoparticle Probes for Bioimaging

Researchers at the University of Michigan are developing nanoprobes that can be used with magnetic resonance imaging (MRI). Nanoparticles with a magnetic core are attached to a cancer antibody that attracts cancer cells. The nanoparticles are also linked with a dye, which is highly visible on the MRI. The cancer cells can then be destroyed by laser or low-dosage killing agents that attack only the diseased cells. In brain tumours, the presence of cancer can weaken the blood-brain barrier. Research is being conducted to have nanoprobes cross the weakened barrier to the tumour but not cross into healthy brain tissue.

Another group at Washington University is using nanoparticles to attracts to proteins emitted from newly forming capillaries that deliver blood to solid tumours. The nanoparticles circulate to the bloodstream and attach to the blood vessels containing their complementary protein. Once attached, chemotherapy is released into the capillary membrane. The nanoparticles travelling in the bloodstream would be able to locate additional cancer sites which may have spread to other parts of the body. Nanotechnology can also be used to improve the contrast agents for MRIs using fullerenes to encapsulate contrast agents such as gadolinium. Luna innovations claims that traditional images can be enhanced by 50 times. If successful, this can result in the use of lower magnetic fields through smaller, cheaper and portable imaging equipment. Alternatively, smaller amounts of contrast agents could be used which reduce the risk of allergic reaction in some patients.

CONCEPT CHECK

1. Explain the role of QD in FACS.
2. How can DNA be conjugated with nanoparticles?
3. Mention the name of the commercially available nano drug for Hepatitis C.
4. Why are micro-fabricated sensors used as central tools in the development of many types of sensors?
5. What is the main role of photodynamic therapy?

MIND TWISTER

1. How do nanoparticles escape from phagocytic attack?

REFERENCES

Ballerstadt, R. and Schultz, J.S. (2000). A fluorescence affinity hollow fiber sensor for continuous transdermal glucose monitoring. *Anal. Chem.* 72, 4185–4192.

Chen, Q., Huang, Z., Chen, H., Shapiro, H., Beckers, J. and Hetzel, F. W. (2002). Improvement of tumor response by manipulation of tumor oxygenation during photodynamic therapy. *Photochem. Photobiol.* 76, 197–203.

Foster, T.H., Hartley, D.F., Nichols, M.G. and Hilf, R. (1993). Fluence rate effects in photodynamic therapy of multicell tumor spheroids. *Cancer Res.* 53, 1249–1254.

Garcia-Zuazaga, J., Cooper K.D. and Baron E.D. (2005). Photodynamic therapy in dermatology: Current concepts in the treatment of skin cancer. *Expert Rev. Anticancer Ther.* 5, 791–800.

Grayson, A.C.R., Shawgo, R.S., Li, Y. and Cima, M.J. (2004). Electronic MEMS for triggered delivery. *Adv. Drug Del. Rev.* 56: 173 Y184.

Henderson, B.W., Busch T.M. and Snyder. J.W. (2006). Fluence rate as a modulator of PDT mechanisms. *Lasers Surg. Med.* 38, 489–493.

Lanza, G.M., Wallace, K.D., Scott, M.J. *et al.* (1996). A novel site-targeted ultrasonic contrast agent with broad biomedical application. *Circulation.* 95: 3334–3340.

Makrides, S.C., Gasbarro, C. and Bello, J.M. (2005). Bioconjugation of quantum dot luminescent probes for western blot analysis. *Biotechniques.* 39, 501–506.

Martin, F., Walczak, R., Boiarski, A., Cohen, M., West, T. Cosentino, C. and Ferrari., M. (2005). Tailoring width of microfabricated nanochannels to solute size can be used to control diffusion kinetics. *J. Control Release.* 102: 123 Y133.

Maruccio, G., Visconti, P., Arima, V., Amico, S.D., Basco, A., Eliana D'Amone, Cingolani, R., Rinaldi, R., Masiero, S., Giorgi, T. and Gottarelli. G. (2003). Fabrication of sub-10 nm planar nanotips for transports experiments of biomolecules. *Nano Lett.* 3, 479.

Pantarotto, D., Partidos, C.D., Graff, R., Hoebeke, J., Briand, J., Prato, M. and Bianco, A. (2003). Immunization with Peptide-Functional Carbon Nanotubes Enhances Virus-specific Neutralizing Antibody Responses. *J. Am. Chem. Soc.* 125, 6160.

Tokumasu, F. and Dvorak, J. (2003). Development and application of quantum dots for immunocytochemistry of human erythrocytes. *J. Microsc.* 211, 256–261.

Yamazaki, T., Kojima, K. and Sode, K. (2000). Extended-range glucose sensor employing engineered glucose dehydrogenases. *Anal. Chem.* 72, 4689–4693.

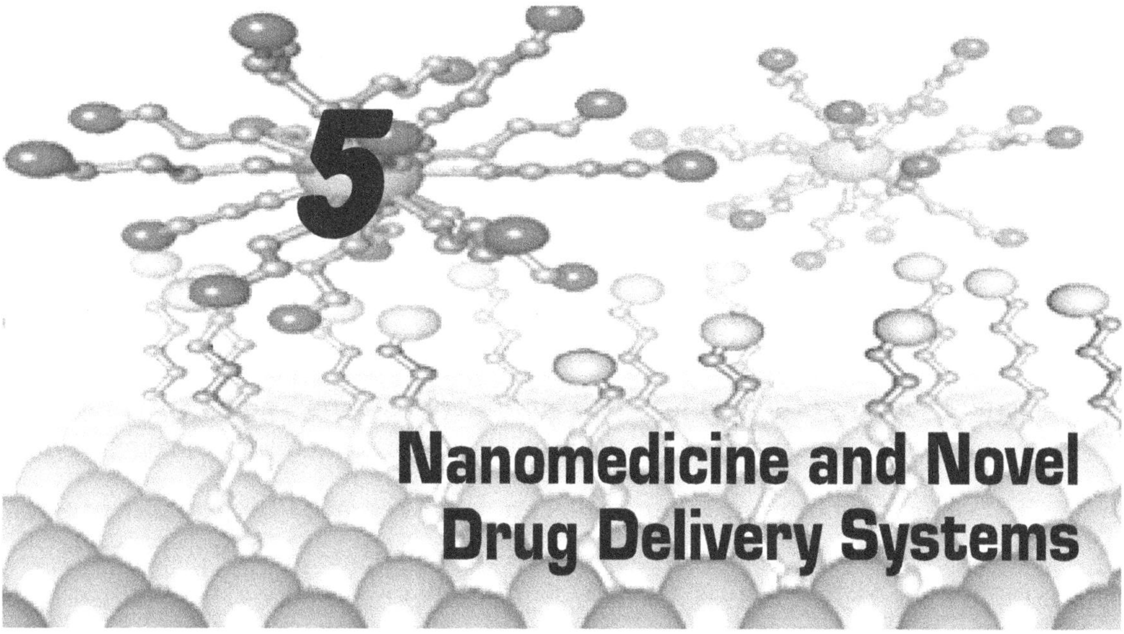

5 Nanomedicine and Novel Drug Delivery Systems

INTRODUCTION

Nanotechnology is a new direction in science and technology, which is intensively developed during the last decade and represents one of the most important directions in the technological developments of the leading countries in 21st century. By analogy with existing microtechnologies, the nanotechnologies may be considered as the technologies operating with nanometre objects. These include new nanomaterials: nanotubes, fullerenes, nanocomposites, porous materials, ultradispersed powders, photon crystals, supramolecular ensembles and constructs, thin films and superficial layers, micellar systems and microemulsions, liquid crystals, liposomes, biomembranes, etc. Employment of nanotechnologies and nanomaterials opens new perspectives in electronics, chemical industry, energetics, biology and medicine, etc. Analytics believe that the world market of nanodevices and modelling at the molecular level will approximately increase at a rate of 28% per year.

Today nanotechnologies significantly influence all aspects of our life. Their commercial use involves all branches of industry, medicine, agriculture, etc. Transition from the "micro" to the "nano" represents qualitative rather than quantitative transition from manipulation of the matter to controlled manipulation by individual atoms and molecules. Due to their sizes the nanoparticles acquire new physico-chemical properties and functions, which significantly differ from properties and functions of atoms and molecules constituting large-sized particles.

Nanobiotechnology or biomolecular nanotechnology is the term coined for biological application or use of nanotechnologies. Basically, the nanotechnologies represent a link between alive and lifeless nature. The study of structure and functions of natural nanoconstructions is the necessary step in creation of nanobiodevices. The major goal of nanobiotechnologies consists in understanding of functioning of biological units for design of small components of a living matter by means of special materials and interfaces. Thus, nanobiotechnologies promote tight cooperation of life sciences with physics, chemistry and engineering.

Medical application of nanobiotechnologies resulted in appearance of a new field. According to the definition given by Freitas: "Nanomedicine is the comprehensive monitoring, repair, construction and control of human biological system, working from the molecular level using engineered nanodevices and nanostructures." In medicine the nanobiotechnological capacities serve for control of physical, chemical, and biological processes in living organisms by nanomaterials and nanoparticles operating at the molecular level. Now nanotechnologies are used for design of nanodevices, which may carry out various operations/ manipulations from diagnostics and monitoring human biological system.

DRUG DELIVERY SYSTEMS

A wide variety of polymeric biomaterials are components of the various formulations and devices that are routinely used for delivering drugs to the body. The current state-of-art of drug delivery systems (DDS) is that they release drugs in a controlled manner. The major advantage of developing systems that release drugs in a controlled manner can be predicted in blood plasma levels following a single dose administration of a therapeutic agent. Successful efforts to produce pharmaceutical formulations that would prolong the action of therapeutic agents for any particular dose, thus minimizing the frequency between dosing, date back to the late 1940s and early 1950s with the introduction of the first commercial product, known as Spansules®. This product was designed to increase the duration of orally administered drugs and consisted of small drug spheres coated with a soluble coating. By using coatings with varying thickness, capsule dissolution times could be varied, thus prolonging the action of one dose of the therapeutic agent. Such formulations are now known as "sustained release or prolonged release products," and more recently they have been called "controlled-delivery" products.

However, the functionality of such products depends greatly on the local *in vivo* environment, and as such varies from patient to patient. For this reason, a major effort has been underway since the late 1960s focused on the development of products that are capable of releasing drugs by reproducible and predictable kinetics. Ideally, such products are not significantly affected by the local *in vivo* environment so that patient-to-patient variability is greatly reduced. Such devices are known as "controlled-release" products. Controlled-release polymeric systems can be classified according to the mechanism that controls the release of the therapeutic agent.

Microcapsules and Microspheres

A microcapsule includes a well-defined core containing the therapeutic agent, and a well-defined polymer membrane surrounding the core. In controlled drug delivery applications, such systems are of interest because drug release takes place by diffusion from the core through the rate-limiting membrane. Microcapsules may be prepared by a complex, three-phase colloidal dispersion method. They may also be made by coating microspheres, which is a simpler process. The attractiveness of achieving long-term zero-order drug release with microcapsules may be offset by fairly expensive manufacturing methods that limit the use of microcapsules to human therapeutics that are not price-constrained and the possibility of membrane rupture and abrupt release of the core material with the consequent overdosing potential. On the other hand, microspheres are most often prepared by a simpler microemulsion polymerization technique. In microspheres, there is no outer membrane and such devices are in effect matrix devices with the drug more or less uniformly distributed in the polymer matrix. Release kinetics are of diffusion from a matrix system, and particle size distribution will be an important variable.

Microspheres and microcapsules are special categories of microparticulates. "Coarse" microparticles may be prepared by simple phase separation and lyophilization or solid diminution techniques, such as grinding and micronizing, using polymer/drug solutions, and size distributions will be broad, sometimes necessitating sieving.

The Enhanced Permeability and Retention Effect (EPR)

The principle rationale driving the development of polymer–drug conjugates, as well as micellar drug delivery systems is the enhanced permeability and retention (EPR) effect, first described by Maeda and co-workers. This effect was attributed to the combination of two effects, 1) the disorganized pathology of tumour vasculature with a discontinuous endothelium which makes it hyperpermeable ("leaky") to circulating macromolecules and to particulates such as liposomes or micelles having an average diameter between 50 and 100 nanometres, plus 2) poor lymphatic drainage, resulting in the retention of the delivery vehicle in the interstitial space of the tumour. Such materials, when administered intravenously, will passively accumulate in tumors and can result in an intra-tumour concentration as much as 70-fold over that found systematically. Polymeric micelles and liposomes may similarly accumulate within the interstitial space of a tumour due to EPR. This type of DDS has been called "passive targeting".

PEG-Protein Conjugates

Proteins are now recognized as an important therapeutic entity, but there are significant problems associated with their use. Dominant among these is antigenicity and proteolytic degradation, and both of these can result in very short plasma half-lives. These problems can, to some extent, be alleviated by conjugation with poly(ethylene glycol) (PEG), a process known as PEGylation, first described by Davis, Aubuchowski and colleagues.

The exact mechanism by which PEG protects proteins from enzymatic degradation, reduces antigenicity and prolongs circulation time is not entirely understood. However, because the PEG chains bind and retain water (about 2–3 water molecules per ether oxygen group), it is likely that the PEG chains form a protective "cloud" at the protein surface that strongly retains hydration water and thus resist closer approach to the protein itself. This protective cloud then shields the protein from the action of degradation enzymes and reduces interaction with receptors, thus reducing immunogenicity. Further, it may also block recognition and capture by the reticuloendothelial system (RES) and further enhance circulation times. Since protein drugs have an active site, this protective cloud can also shield the active site and reduce biological activity. However, it has been suggested that the increased circulation time more than compensates for the reduced activity.

POLYMER THERAPEUTICS

Polymer—Drug Conjugates

An effective polymer-drug conjugate is stable while in blood plasma, but then gradually releases the drug once it is localized by the EPR effect within tumour, and the drug is then taken up by the tumour cells. One of the most extensively investigated polymers used in polymer–drug conjugates is polyN-(2-hydroxypropyl) methacrylamide, @known as HPMA. In this particular polymer system, the drug is covalently attached to the water-soluble polymer backbone via a glycine-phenylalanine-leucine-glycine spacer. This spacer is stable in circulation, but once it has been internalized by the tumour cell via endocytic uptake, the enzyme cathepsin B cleaves the peptide spacer within the lysosome, and the drug is liberated from the polymer conjugate and diffuses from the lysosome to the nucleus where it can kill the cell. This has been called "lysosomotropic" drug delivery.

Polymeric Micelles

Amphiphilic, A-B, block copolymers, i.e., block copolymers constructed from a hydrophobic segment, are known to assemble in an aqueous environment into polymeric micelles. This self assembly is due to the large difference in solubility between the two segments. Polymeric micelles have a fairly narrow size distribution and have a unique core-shell structure where the hydrophobic segments form an inner core surrounded by the hydrophilic segments, known as "corona". Polymeric micelle only exist above the critical micelle concentration (CMC). Below that concentration the micelles dissociate into the individual polymer molecules, known as unimers.

Polymeric micelles have excellent potential as DDS. Among these is their ability to solubilize poorly water-soluble drugs and thus increase drug bioavailability. They can also passively target tumours due to the EPR effect. Like liposomes, micelles can also be targeted by attachment of specific ligands at the ends of the hydrophilic block segment. Although other hydrophilic polymers can be used to solubilize and stabilize hydrophobic drugs.

Liposomes

The first observation that ordered structures are obtained when water-insoluble lipids, such as phospholipids, are mixed with an excess of water was made by

Bangham and co-workers in the 1960s. These ordered structures eventually arrange into concentric, closed spherical membranes known as liposomes or vesicles. Such liposomes can consist of one or a multiplicity of bilayer membranes. When the diameter of such vesicles are about 25–100 nm, they are known as small unilamellar vesicles (SUV) and when the diameter is between about 100 nm to many microns, they are known as multilameller vesicles (MLV).

Liposomes are able to entrap water-soluble solutes in the aqueous inner core. Lipophilic drugs may be entrapped in the lipid bilayers by combining such drug with the phospholipids used during their preparation. Vesicles larger than about 200 nm are rapidly cleared from the blood and end up with the macrophages of the reticuloendothelial system (RES). Circulation times of liposomes can be augmented by coating with PEG or sialic surfactants. In the former case PEG is conjugated with the lipid tail, which inserts into the liposome's lipid bilayer—"PEGylation" of liposomes. This polymer forms a water-retaining coating around the liposome and prevents recognition by the RES system. Such liposomes have been termed as "stealth liposomes."

While passive targeting of liposomes to the RES is of some interest, a more interesting approach is one where liposomes with target recognition properties are constructed. Such "active targeting" requires the identification of suitable receptors on the surface of the liposome that can recognize this target. When the liposome is PEGylated, the targeting ligand is conjugated to the outer end of the PEG molecules.

NANOMATERIALS FOR DRUG DELIVERY ACROSS THE BLOOD BRAIN BARRIER (BBB)

The BBB is the most restrictive barrier in the body, which prevents most small molecules and nearly all macromolecules from entering the CNS. Current strategies used for drug delivery to the brain include invasive delivery, temporary disruption of the BBB, as well as the use of drug-delivery systems. While direct injection in selected cases can be an effective invasive modality for local delivery (e.g. in some tumours), it is not efficient for brain metastasis or neurodegenerative diseases, which require therapeutic agents to be widely spread in the brain. Reversible opening of the BBB by an osmotic or chemical method allows therapeutic agents to enter the brain. However, this approach can also result in significant damage to the brain. In comparison, selective delivery of diagnostic and therapeutic molecules to the whole brain through the vascular route using drug-delivery system is much less invasive and is potentially safe. This strategy has been receiving significant attention with remarkable development of nanomedicine.

This section presents a brief overview of these technologies including liposomes, nanoparticles, polymeric micelles, nanogels, and dendrimers for CNS drug delivery. Liposomes are, perhaps, the earliest type of nanomaterial developed for drug delivery. They are vesicles composed of one (unilamellar) or several (multilamellar) lipid bilayers surrounding internal aqueous compartments. Relatively large amounts of drug molecules can be incorporated into the aqueous compartment (water-soluble compounds) or lipid bilayers (lipophilic compounds). Conventional liposomes are rapidly cleared from circulation by the reticuloendothelial system (RES). Extended circulation time can be accomplished by decreasing the particle size (<100 nm) and by liposome-surface modification with polyethylene glycol (PEG). To target PEGylated liposomes to the brain, they can be additionally modified (vectorized) with monoclonal antibodies to glial fibrillary acidic proteins, transferring receptors (OX26), or human insulin receptors. Such immunoliposome constructs were successfully used to deliver small drugs, Daunomycin® and Digoxin®, as well as DNA to the brain. Nanoparticles (NPs) are solid colloidal particles often made of insoluble (bio) degradable polymers. As carriers for drug delivery to the brain, NPs need to be small (<100 nm) and stable in the blood, as well as avoid the RES, neutrophil activation, platelet aggregation, and inflammation. Surfactant-coated polybutylcyanoacrylate NPs were shown to successfully deliver analgesics (Dalargin, Loperamide), anti-cancer agents (Doxorubicin), and anticonvulsants (NMDA receptor antagonist, MRZ 2/576) across the BBB. In another work, NPs conjugated with the metal chelator, Desferioxamine,® were reported to cross the BBB and reduce the metal load in neuronal tissue. These results would be of significance for mitigating the harmful effects of oxidative damage in AD and other neurodegenerative diseases. Overall, the mechanism of entry of NPs as well as liposomes into the brain remains unclear and, in some cases, may include disruption of the tight junctions of brain microvessel endothelial cells forming the BBB.

Polymeric micelles are another type of nanomaterial that have attracted considerable attention as carriers for drug delivery and diagnostic imaging agents. These micelles form spontaneously in aqueous solutions of amphiphilic block copolymers and have core-shell architecture. The core is composed of hydrophobic polymer blocks [e.g. poly(propylene glycol) (PPG), poly(D,L-lactide), poly(caprolactone), etc.] and a shell of hydrophilic polymer blocks (often PEG). The size of polymeric micelles usually varies from ca. 10 to 100 nm. Their core can incorporate considerable amounts (up to 20–30% weight) of water-insoluble drugs preventing premature drug release and degradation. The shell stabilizes micelles in dispersion and masks the drug from interactions with serum proteins and untargeted cells. After reaching target cells, the drug is released from the micelle via diffusion. Several clinical trials are completed or are underway to evaluate polymeric micelles for delivery of anti-cancer drugs.

One study has shown that polymeric micelles of Pluronic block copolymers, after conjugation with an antibody against $\alpha 2$-glycoprotein or insulin, showed increased delivery of a drug (haloperidol) or a fluorescent probe to the brain *in vivo*. Pluronic block copolymers contain two hydrophilic PEG and one hydrophobic PPG blocks (PEG-PPG-PEG). They were shown to bind with the membranes of brain microvessel endothelial cells and to inhibit a drug efflux transport protein, p-glycoprotein (Pgp) that severely restricts transport of many drugs to the brain. As a result of formulating Pgp-dependent drugs with Pluronic, the bioavailability of such drugs to the brain was increased. Another recent work used pluronic molecules capable of crossing the membranes of brain microvessel endothelial cells to increase delivery of proteins to the brain. It was shown that by conjugating pluronic molecules with a model protein, horseradish peroxidase (HRP), the transport of HRP across the BBB *in vitro* and *in vivo* was considerably enhanced. Altogether, significant promise has been achieved by utilizing Pluronic block copolymers and polymeric micelle systems for CNS drug delivery, and one should expect further research and development in this direction.

Polyion complex micelles (also termed "block ionomer complexes") are novel nanosystems for incorporation of charged molecules. They are formed as a result of the reaction of double hydrophilic block copolymers containing ionic and nonionic blocks with macromolecules of opposite charge including oligonucleotides, plasmid DNA and proteins, or surfactants of opposite charge. For example, block ionomer complexes were prepared by reacting trypsin or lysozyme (that are positively charged under physiological conditions) with an anionic block copolymer, PEG-poly(α, β-aspartic acid). Such complexes spontaneously assemble into nanosized particles having core-shell architecture. The core contains polyion complexes of a biomacromolecule and an ionic block of the copolymer. The shell is formed by the nonionic block. In the case of surfactant-based complexes, the core is composed of mutually neutralized surfactant ions and polyion chains. It contains hydrophobic domains of surfactant tail groups and can additionally incorporate water-insoluble drugs. Depending on surfactant and block copolymer architectures, the complexes assume different morphologies including vesicles and micelles of different shapes. These nanomaterials are versatile and can incorporate solutes of different structures with a high loading capacity.

Furthermore, they can release solutes upon change of environmental conditions such as pH (acidification), concentration, and chemical structure of elementary salt. These nanomaterials were shown to efficiently deliver DNA molecules *in vitro* and *in vivo*, and were used to stabilize enzymes for cell-mediated drug delivery discussed in the next section. Nanogels are another novel class of nanomaterials proposed for drug delivery. They represent hydrogels of cross-linked polymer networks that often combine ionic and nonionic chains. Such networks can incorporate charged molecules such as siRNA, DNA, oligonucleotides, and low-molecular mass compounds, which bind to oppositely charged ionic chains. Nanogels were shown to enhance transport of incorporated oligonucleotides to the brain *in vitro* and *in vivo*. The mechanism of nanogel-mediated delivery of oligonucleotides apparently involves transcytosis across brain microvessel endothelial cells. The permeability of oligonucleotides with nanogels was enhanced when the nanogel surface was modified with polypeptides (transferrin or insulin) that bind receptors at the luminal side of the brain microvessel endothelial cells and transport to their abluminal side. Nanogels are currently being investigated for CNS delivery of various low molecular mass compounds and biomacromolecules including nucleoside analogs and plasmid DNA.

Finally, many studies have focused on using dendrimers as carriers of small drugs and biomacromolecules. Dendrimers are repeatedly branched polymer molecules containing cascade of branches grown from one or several cores. They contain three architectural domains: 1) the core, to which the branches are attached, 2) the shell of the branches surrounding the core, and 3) the multivalent surface formed by the branches termini. Compared to most other nanomaterials described in this section, dendrimers have a smaller size and lower polydispersity. A typical dendrimer molecule, for example poly(amidoamine) (PAMAM) dendrimer, has a diameter ranging from 1.5 to 14.5 nm. Various solutes can be entrapped within the dendrimer interior cavities (dendritic boxes) during the dendrimer synthesis. In this case, a dense shell can prevent diffusion of the solutes from the interior, even after prolonged heating, solvent extraction, or sonication. Dendritic boxes may be suitable for drug delivery if a shell can be degraded under physiological conditions resulting in drug release. Dendritic structures can also be used as building blocks for attachment of various functional moieties to the surface. For example, hydrophobic and hydrophilic polymer blocks can be grafted to the surface resulting in formation of unimolecular micelles. Unlike conventional micelles, such unimolecular micelles are stable upon dilution and do not disintegrate during circulation in the body.

Several studies evaluated dendrimers for CNS intratumoral delivery of dendrimer conjugates with anti-cancer agents to treat glioma. Transport of dendrimers across cell barriers was evaluated using intestinal epithelial cells (Caco-2;). Notably, the generation and surface properties of dendrimers were found to be very important. Cationic dendrimers were generally more toxic and disrupted the tight junctions. These effects increased as dendrimer generation and consequently, net surface area increased. Surface modification of dendrimers with carboxylic groups greatly decreased the toxicity, although the modified dendrimers still opened tight junctions. It is also well known that the fourth- and fifth-generation PAMAM dendrimers can bind DNA and enhance DNA delivery into a cell. It has recently been demonstrated that Starburst™ PAMAM dendrimer/DNA complexes can also penetrate the BBB resulting in enhanced expression of a reporter gene in the brain. Gene transfer into brain capillary cells have been also shown using a transferrin-conjugated PEG modified PAMAM dendrimer. A widespread expression of an exogenous gene in mouse brain was observed after i.v. administration of complexes of this dendrimer and DNA, suggesting potential use of such materials as nonviral vectors for CNS gene delivery. Clearly, the use of polymers and nanomaterials for drug delivery to the brain is still in its infancy.

However, one should expect new developments in this area, which ultimately can revolutionize the diagnosis and therapy of neurodegenerative disorders. Notably, several nanoscale drugs (nanomedicines), such as liposomal Doxorubicin, Doxil® or albumin-bound Paclitaxel, Abraxane® have been used in clinic already for treatment of cancer and other diseases. Others, such as Doxorubicin incorporated in polymer micelles, SP1049C and NK911, are in clinical trials. Most importantly, such studies have demonstrated that nanomaterials can be safely administered in the human body. Clearly, some of the approaches developed in these earlier studies can and will be extended to develop efficient therapeutic modalities for neurodegenerative disorders.

NANOPARTICLES TARGET CANCER CELLS *IN VIVO*

Using nanoparticles (NPs) for drug delivery of anticancer agents has significant advantages such as the ability to target specific locations in the body, the reduction of the overall quantity of drug used, and the potential to reduce the concentration of the drug at nontarget sites resulting in fewer unpleasant side effects. For intravenously injected particles, biocompatibility is the first issue. The second is to promote specific carrier–target interactions or devise specific carrier physicochemical properties to enable selective targeting. Recent research has made advances on both fronts. Researchers from Harvard Medical School, Massachusetts Institute of Technology (MIT), MIT-Harvard Centre for Cancer Nanotechnology Excellence, and Gwangju Institute of Science and Technology in Korea have reported the successful use of targeted NP-aptamer bioconjugates for *in vivo* chemotherapy. The aim of the joint effort of Omid C. Farokhzad and Robert Langer's groups was to design a system able to target cancer cells specifically, introduce chemotherapy drugs directly into the cells and release them over time, and avoid uptake by noncancerous cells. To achieve this, the researchers devised an NP-based system using biodegradable and biocompatible poly(D,L-lactic-co-glycolic acid).

An anticancer drug, in this case docetaxel (Dtxl®), is encapsulated in the 150 nm NPs. The surface of the NPs is then functionalized with aptamers (Apt)—nucleic acid ligands—that promote cells-specific uptake by binding to target antigens, analogous to antibodies. In this case, the RNA aptamers that recognize the prostate-specific membrane antigen (PSMA), which is expressed by prostate cancer cells, are used. The NPs are also functionalized with poly(ethylene glycol) (PEG) to reduce uptake by tissue macrophages and nontargeted cells. The polymeric NPs dissolve slowly once inside the cancerous cell, releasing the Dtxl®. The rate of dissolution can be designed into the system so that drugs are released over a predetermined time. After a single administration of the Dtxl®-NP-Apt bioconjugates into mice, which had been implanted with human prostate cancer cell lines, five out of seven animals showed complete tumour reduction and all survived the 109-day study. The ability to treat tumours in one go is particularly attractive. "Early diagnosis of cancer creates the opportunity for localized treatment options," explains Farokhzad. Since all of the components of the NP-based system, except the RNA aptamers, are already approved by the Food and Drug Administration for clinical use, Farokhzad is confident that such NP-based treatment systems could be feasible in the near future, not only for cancer but other diseases as well. As Langer explains, "[we] can put different things inside or on the outside of the NPs. This technology can be applied to almost any disease."

Dendrimers as Nanoparticulate Drug Carriers

The discovery and creation of new drugs is a timely and costly process. It is estimated that every new drug takes 12–15 years to develop at a cost of over $ 800 million. A more efficient approach would be the devising of effective drug delivery systems for already developed experimental drugs that failed to make it to the market. Controlled release systems can improve the effectiveness of drug delivery by sustained release of the compound over time or by release at a specific target. By controlling the time and location of delivery, side effects can be minimized and drug efficacy can be maximized, thus leading to a lower dosage for patients.

Currently the two common drug delivery systems are liposomes and polymeric systems. These both have limited applications, liposome-based systems have poor stability and difficulty targeting specific tissues, and linear polymers are polydisperse. Dendrimers offer advantages including a lower polydispersity index, multiple sites of attachment, and a controllable, well-defined size and structure that can be easily modified to change the chemical properties of the system. In addition, macromolecules such as dendrimers have an enhanced permeability and retention effect that allows them to target tumour cells more effectively than small molecules.

Dendrimers were reported by Vögtle at 1978 by cascade method, but the synthesis method was clearly explained by D.A. Tomalia at 1985 by divergent method. Frechet (1990), has given another method called convergent, for the synthesis of dendrimers, even though most researchers follow only the divergent method, to construct the dendrimeric structure. Dendrimers have applications in gene and antisense therapy, magnetic resonance imaging and in boron neutron capture therapy. Advances in dendrimer delivery systems, biodegradable dendrimers and release from dendrimers can be applied to drug delivery in addition to other applications.

DENDRIMERS AS DRUG CARRIERS

Two methods of dendrimer drug delivery are encapsulation of drugs and dendrimer–drug conjugates. Encapsulation of drugs uses the steric bulk of the exterior of the dendrimer or interactions between the dendrimer and the drug to trap the drug inside the dendrimer. Dendrimer–drug conjugates have the drug attached to the exterior of the dendrimer. Most of these conjugates are prodrugs and are inactive or have decreased activity relative to the free drug.

Encapsulation of Drugs in Dendrimers

The flexible branches of a dendrimer, when constructed appropriately, can provide a tailored sanctuary containing voids that provide a refuge from the outside environment wherein drug molecules can be physically trapped. Encapsulation of hydrophilic, hydrophobic or even amphiphilic compounds as guest molecules within a dendrimer can be enhanced by providing various degrees of multiple hydrogen-bonding sites or ionic interactions or highly hydrophobic interior void spaces. A wide variety of molecules have been successfully encapsulated inside dendrimers. In early experiments, compounds used to demonstrate the "guest molecule" concept included easy-to-visualize dye molecules such as rose Bengal and Reichardt's dye as well as pyridine and peptides. More recently, actual drugs, including 5-fluorouracil, 5-amino salicylic acid, pyridine, mefanminic acid and diclofenac, paclitaxel, docetaxel, as well as the anticancer agent 10-hydroxycamptothecin have been successfully encapsulated. Together, these results demonstrate that encapsulation is a general strategy for the delivery of low molecular weight compounds by dendrimers. This method is anticipated to be of particular value when display of the bioactive molecule on the surface of the dendrimer induces unwanted immunogenicity or reduces biocompatibility.

Conjugation of Drug in Dendrimers

The strategy of coupling small molecules to polymeric scaffolds by covalent linkages to improve their pharmacological properties has been under experimental test for over three decades. These conjugates range from small molecules drugs such as ibuprofen, fluorescent and radioactive imaging agents, oligonucleotides, oligosaccharides and peptides as well as much larger molecules such as monoclonal antibodies. Similar to encapsulated guest molecules that generally require from the void spaces of a dendrimer to gain bioactivity, a covalently delivered dendritic conjugate must also be cleaved within the target cell to regenerate the active cytotoxic agent. At the same time, to ensure systemic non-toxicity, the covalent linker must be stable in circulation. Though conjugation serves to high drug payload, it could also result in limited aqueous solubility of the drug–dendrimer conjugate, especially for hydrophobic drugs. In addition, the conjugation of drug to polymer is a time-consuming process, requiring additional synthetic steps and could sometimes result in decreased therapeutic benefit principally due to the failure of conversion or slow conversion to the parent drug or decreased permeability of the conjugate.

Thus, physical encapsulation seems to be a more effective and attractive approach for delivery of small guest molecules. The encapsulation of drug in the dendritic voids protects labile molecules from degradation and also eliminates solubility issues associated with poorly soluble compounds. However, it is reported that drugs loaded non-covalently into dendrimers are released rapidly and rendering them ineffective for targeted delivery. Hence presently, improvements in the encapsulation efficacy and release properties remain as the major challenges in dendrimers-based delivery systems.

Cell-Penetrating Peptides (CPP) in Combination with Nanoparticles for Novel Drug Delivery

The plasma membrane of mammalian cells widely restricts the passage of large, charged and hydrophilic compounds into cells, preventing most peptide, protein and nucleic acid biopharmaceuticals to reach intracellular targets, and frequently results in low or zero biological efficacy. The architecture of the plasma membrane of polarized epithelial cells is well characterized. Epithelial cells are connected by a junctional complex encircling the apex of each cell. Tight junctions are the most apical structures of the junctional complex which also includes adherens junctions, desmosomes and gap junctions. Tight junctions constitute a barrier to the paracellular diffusion of solutes from the lumen to the tissue parenchyma (gate function) and restrict the lateral exchange of lipids and proteins between the apical and basolateral plasma membrane domains (fence function).

Cell-penetrating peptides (CPPs) constitute a collection of various families of short peptide sequences that have been shown to translocate across the plasma membrane via different mechanisms. The prospects of chemical ligations or physical assemblies between CPPs and therapeutic cargoes has, therefore, attracted considerable interest, and created widespread hopes to exploit the CPP approach for drug delivery purposes, gene therapy and vaccine development. Various oligocationic cell-penetrating peptides, e.g. HIV-1 derived Tat peptides, penetratin (pAntp), oligoarginine peptides or the weakly cationic, human calcitonin-derived CPPs, etc. They were commonly considered for the therapeutic delivery of peptides, proteins, oligonucleotides, plasmids, peptide nucleic acids (PNAs) and even nanoparticles. N-terminally carboxyfluorescein/CF-labelled, and oligocationic CPPs are the two recently developed molecular tools for cellular entry. One is a branched derivative of the linear human calcitonin denoted as CF-hCT(9-32), that carries an oligocationic, SV40-derived nuclear localization sequence, the GPDEVKRKKKP motif, in the form of a side-branch to the main peptide chain. The other one is the proline-rich sweet arrow peptide CF-SAP, a linear trimer of repetitive VRLPPP domains, and conceived as an amphipatheic version of a polyproline sequence related to g-zein, a storage protein of maize. The internalization of both CPPs into HeLa cells was recently found to follow the pathway of lipid raft-mediated endocytosis prior to endosomal escape.

Diverse mammalian cell lines have been used to study the internalization of CPPs across the plasma membrane. In the majority of cases, "leaky" cell models, such as HeLa, KB 3-1, Bowes human melanoma or MC57 fibrosarcoma cells, have been used for CPP internalization. When cultured on solid surfaces, such cell lines lack the ability to form tight junctions and form cell layers of high permeability. Cellular models with junctional complexes bear more relevance for therapeutic considerations, namely for the delivery of CPP-cargo constructs into epithelia. Cell line dependence of CPP internalization has been suggested by different research groups.

Altogether, some research findings emphasize the importance of relevant cell models to study CPP internalization and point towards inflamed epithelia as potential niche for CPPs as drug delivery vehicles.

BIORESPONSIVE HYDROGELS

Modern biomaterials are increasingly designed to interface with biological tissues in predefined ways. An important class of biomaterials are those that are highly hydrated-hydrogels. Hydrogels consist of elastic networks with interstitial spaces that contain as much as 90–99% w/w water. They are prepared by chemical polymerization or by physical self assembly of manmade or naturally occurring building blocks. Most commonly, these building blocks are macromolecules in a variety of architectures, including cross-linked polymers, entangled fibrillar networks or colloidal assemblies. Recently, small amphiphilic molecules have emerged as a new class of hydrogelators, forming supramolecular or molecular hydrogels.

Hydrogel materials are increasingly studied for applications in biological sensing, drug delivery and tissue regeneration for a number of reasons:

- Hydrogels provide suitable semiwet, three-dimensional environments for molecular-level biological interactions
- Many hydrogels provide inert surfaces that prevent nonspecific adsorption of proteins, a property known as antifouling
- Biological molecules can be covalently incorporated into hydrogel structures using a range of well-established chemistries
- Hydrogel mechanical properties are highly tunable, for example elasticity can be tailored by modifying cross-link densities
- Hydrogels can be designed to charge properties (e.g. swelling/collapse or solution-to-gel transitions) in response to externally applied triggers, such as temperature, ionic strength, solvent polarity, electric/magnetic field, light or small (bio)molecules.

In tissue regeneration, biomaterials may be designed to contain precisely positioned bioactive ligands that instruct cell behaviour. A well known example is the cell-adhesive tripeptide, Arg-Gly-Asp (RGD). This peptide is derived from fibronectin, a component of the extracellular matrix (ECM) to which cells attach *in vivo*. Incorporation of RGD into hydrogel structures instructs many cell types to attach to the material via cell-surface proteins called integrins. Three types of stimuli for bioresponsive hydrogel systems can be distinguished.

First, hydrogel materials can be modified to contain small biomolecules that selectively bind to biomacromolecules, including protein receptors or antibodies. Upon binding, a macroscopic transition follows such as, swelling/collapse or solution-to-gel transitions.

Second, systems may be modified with the enzyme-sensitive substrate. Since enzymes are highly selective, materials can be programmed to respond to a specific enzyme by incorporation of the specific substrate (or a substrate mimic).

This concept is especially attractive because the distribution of enzymes can differ between healthy and diseased cells, between different cell types and during cellular migration, differentiation and cell division.

Third, systems may have biomacromolecules, such as enzymes, incorporated into their structures that recognize small biomolecules. Enzymatic conversion of these biomolecules into molecules with different physical properties (e.g. an acid or basic compound) then triggers hydrogel swelling or collapse.

Bioresponsive Hydrogels for Drug Delivery

Much work on bioresponsive hydrogels for drug delivery relates to the release of insulin in response to raised blood sugar levels as a potential autonomous treatment of insulin-dependent diabetes. In one approach, glucose oxidase molecules are immobilized onto a basic polymeric carrier. Following the enzyme reaction that converts glucose to gluconic acid, thereby temporarily lowering the pH, the basic groups on the polymer are protonated, inducing swelling and enhancing the release profile of insulin. This system works as feedback loop: upon release of insulin the sugar levels drop, resulting in a pH increase that stops the release of further insulin.

Researchers in Texas have functionalized hydrogel nanoparticles to improve the delivery of orally administered drugs to the small intestines, a site where therapeutic compounds are absorbed into the bloodstream. The team grafted methacrylic acid and ethylene glycol together to create a unique polymer hydrogel that can survive the harsh acidic conditions of the stomach. The hydrogel nanoparticles were also decorated with wheat germ agglutinin (WGA) to make them stick to the intestinal tissue. *In vitro* tests indicate that functionalized hydrogel nanoparticles, soaked with insulin are very efficient at loading and retaining the drug. This makes them a good delivery vehicle for insulin, a protein drug used to treat diabetes. When subjected to the pH conditions found in the stomach, the particles retain insulin. On raising the pH to more neutral levels, such as those found in the small intestines, insulin is released. This happens because methacrylic acid contains carboxyl groups which remain protonated at low pH conditions.

At more neutral pH levels, the carboxyl group deprotonates causing the polymer chains to repel each other. As a result, the hydrogel swells, releasing the drug molecules from the nanoparticle into the system. The hydrogel spheres are designed to be greater than 200 nm in diameter to prevent their uptake by lymphoid cells called Peyer's patches that are also present in the small intestines. Instead the insulin cargo is released directly into the bloodstream, where it is needed. He also explained the reasoning behind using WGA as anchors is that "Functionalization of PEG chains with WGA allows for specific binding to carbohydrate moieties present in the intestinal mucosa to improve residence time of the carrier at the delivery site."

BIODEGRADABLE pH-SENSITIVE MICELLE SYSTEM FOR TARGETING ACIDIC SOLID TUMOURS

Biodegradable pH-responsive drug carriers have the potential not only to provide selective drug release at specific targets like tumours, but also to be rapidly cleared or degraded in the body after delivering the cargo. In recent years, pH-sensitive polymeric carriers in various forms of micelles, vesicles, and nanoparticles have seen rapid development. Particularly, polymeric micelles are being extensively studied as a promising nanoscale drug carrier since the pioneering work in early 1990s. Polymeric micelles have many advantages such as small size (10–200 nm) for passive accumulation in solid tumours by enhanced permeation and retention (EPR), improved stability, biodegradability and high flexibility for structural and chemical modifications.

Early generation polymeric micelles simply accumulated on the tumour extracellular matrix (ECM) and did not provide high-enough concentrations of anticancer drugs to kill the tumours, because most cytotoxic drugs act inside the cells and not on the ECM. Therefore, tumour targeting carriers have been modified such that they accumulate by the EPR effect on tumour cells followed by active internalization into tumour cells. Such improved active targeting technology is being developed by using binding characteristics between tumour-specific antigens and their monoclonal antibodies (mAb), binding fragments specific to a tumour-associated surface antigens, or between ligands and their corresponding receptors. Such active targeting carriers show enhanced capability to translocate the micelles into tumour cells.

In most cases, the therapeutic drugs are released from the polymer scaffold in the lysosomal environment and cleaved by enzymatic degradation or pH sensitive hydrolysis. Even after the drug is released from the carrier, the diffusion through the lysosomal membrane to cytosol and then to the nucleus, has proved to be a formidable barrier. TAT-mediated cytoplasmic uptake of drug conjugates can deliver the cargo directly at the periphery of the nucleus avoiding the endocytotic pathway. The probable mechanism of internalization of TAT (Trans-Activator of Transcription) peptide is via electrostatic interaction and hydrogen-bonding or macropinocytosis depending on the size of the cargo. A major hurdle in using TAT in the body is its non-specificity, since it can interact with any cells. It has been firmly established that the pH of ECM of most solid tumours is relatively lower (pH<7.0) than that of normal tissues. The lower pH comes from the high metabolic rate of tumour cells which leads to production of excess lactic acid and hydrolysis of ATP under hypoxic conditions.

This difference in pH between tumour and normal cells, even though very small, has instigated many investigators to develop pH-sensitive drug carriers that use it as a trigger. Among these, drug carriers having sulphonamide pendant groups are most promising as they have shown sharp transition ranges around physiological pH.

TISSUE REGENERATION/ENGINEERING

The loss or failure of an organ or tissue is a frequent, devastating and costly problem in health care, occurring in millions of patients every year. Furthermore, artificial devices made of plastic, metal, or fabrics are utilized. Mechanical devices such as dialysis machines or total joint replacement prostheses are used, and metabolic products of the lost tissue, such as insulin are supplemented. Although these therapies have saved and improved millions of lives, they remain imperfect solutions.

Tissue engineering represents a new, emerging interdisciplinary field applying a set of tools at the interface of the biomedical and engineering sciences that use living cells or attract endogenous cells to aid tissue formation or regeneration to restore, maintain or improve tissue function. Engineered tissues using the patient's own (autologous) cells or immunologically inactive allogeneic or xenogeneic cells offer the potential to overcome the current problems of replacing lost tissue function and to provide new therapeutic options for diseases such as metabolic deficiencies.

Tissue Engineering using Scaffold Biomaterials

Open systems of cell transplantation with cells being in direct contact with the host organism aim to provide a permanent solution to the replacement of living tissue. The rationale behind the use of open systems is based on empirical observations: dissociated cells tend to reform their original structures when given the appropriate environmental conditions in cell culture–for example, capillary endothelial cells from tubular structures and mammary epithelial cells form acini that secrete milk on the proper substrata *in vitro*. Although isolated cells have the capacity to reform their respective tissue structure, they do so only a limited degree since they have no intrinsic tissue organization and are hindered by the lack of a template to guide restructuring. Moreover, tissue cannot be transplanted in large volumes because diffusion limitations restrict interaction with the host environment for nutrients, gas exchange and elimination of waste products. Therefore, an implanted cell survives poorly more than a few microns from the nearest capillary or other source of nourishment. With these observations in mind, an approach has been developed to regenerate tissue by attaching isolated cells to biomaterials that serve as a guiding structure for initial development. Ideally these scaffold materials are biocompatible, biodegradable into nontoxic products and manufacturable. Natural materials used in this context are usually composed of extracellular matrix components (e.g. collagen, fibrin) or complete decellularized matrices (e.g. heart valves, small intestinal submucosa).

Synthetic polymer materials are advantageous in that their chemistry and material properties (biodegradation profile, microstructure) can be well controlled. The majority of scaffold-based tissue engineering concepts utilize synthetic polymers [e.g. poly(glycolic acid) (PGA), poly(lactic acid) (PLA) or poly(hydroxyl alkonate) (PHA)]. In general, these concepts involve harvesting of the appropriate cell types and expanding them *in vitro*, followed by seeding and culturing them on the polymer matrices. The polymer scaffolds are designed to guide cell organization and growth allowing diffusion of nutrient to the transplanted cells. Ideally, the cell-polymer matrix is prevascularized or would become vascularized as the cell mass expands after implantations. Vascularization could be a natural response to the implant or could be artificially induced by sustained release of angiogenic factors from the polymer scaffold. Since the polymer scaffold is designed to be biodegradable, concerns regarding long-term biocompatibility are obviated. Cells used in tissue engineering may come from a variety of sources including cell lines from the patient themselves (autologous), human donors (allogeneic) or animal sources (xenogeneic). However, allogeneic and xenogeneic tissue may be subjected to immunorejection. Cell-surface modulation offers a possible solution to this problem by deleting immunogenic sites and therefore preventing immunorecognition. A bank of cryopreserved cells would then be possible and genetic engineering techniques could be used to insert genes to replace proteins, such as the LDL receptor or factor IX.

APPLICATION OF MATERIALS IN MEDICINE

Private and public research efforts worldwide are developing nanoproducts aimed at improving health care and advancing medical research. Some of these products have entered the marketplace, more are on the verge of doing so, and others remain more a vision than a reality. The potential for these innovations is enormous, but questions remain about their long-term safety and the risk–benefit characteristics of their usage. Nanoscale materials and devices can be fabricated using either "bottom-up" or "top-down" fabrication approaches. In bottom-up methods, nanomaterials or structures are fabricated from build-up of atoms or molecules in a controlled manner that is regulated by thermodynamic means such as self-assembly. Alternatively, advances in microtechnologies can be used to fabricate nanoscale structures and devices. These techniques, which are collectively referred to as top-down nanofabrication technologies, include photolithography, nanomoulding, dip-pen lithography and nanofluidics. It is perhaps because of the breadth of different approaches in the synthesis and fabrication of nano-molecules and nano-devices that chemical engineers are playing key role in advancing the field of nanotechnology. Nanomaterials and devices provide unique opportunities to advance medicine. The application of nanotechnology to medicine is referred to as "nanomedicine" or "nanobiomedicine" and could impact diagnosis, monitoring, and treatment of diseases as well as control and understanding of biological systems.

Cardiovascular Medical Devices

In no area of medicine have biomaterials played a more critical role in the life-saving treatment of patients than in the cardiovascular system. Blood oxygenators used in cardiopulmonary bypass have made possible open heart surgeries such as coronary artery bypass surgery, valve replacement and repair of congenital or acquired structural cardiac defects. Heart-valve prostheses, both mechanical and bioprosthetic are used to replace dysfunctional natural valves with substantial enhancement of both survival and quality of life. In this situation, the benefit is substantial and has been well documented.

Substitute heart valves The four valves in the human heart play a critical role in assuring the forward blood flow that is critical to proper cardiac function. The tricuspid valve allows flow from the right atrium to the right ventricle, the pulmonary valve from the right ventricle to the pulmonary artery, the mitral valve from the left atrium to the left ventricle and the aortic from the left ventricle to the aorta. Disorders of these valves can causes stenosis (i.e., obstruction to flow), regurgitation (i.e., reverse flow across the valve), or a combination of both stenosis and regurgitation. Some diseases such as infective endocarditis (i.e., infection of a heart value) can cause rapid destruction of the affected valve and can lead to abrupt heart failure and death; others such as degenerative calcific aortic stenosis can take many decades to develop (during which the disease is inapparent) before clinical manifestations appear.

The surgical treatment available for valvular heart disease include repair and valve replacement. Reconstructive procedures to eliminate mitral insufficiency of various etiologies and to minimize the severity of rheumatic mitral stenosis are now highly effective and commonplace, accounting presently for over 70% of mitral valve operations. Repairs are generally preferable, if they can be done. The advantages of repair over replacement relate to the elimination of both the risk of prosthesis-related complications and the need for chronic anticoagulation that is required in many patients with substitute valves, and to a lower rate of post-operative valve-related infection.

Mechanical valves are composed of nonphysiologic biomaterials that employ a rigid, mobile occluder (usually a pyrolytic carbon disk) in a metallic cage (cobalt-chrome or titanium alloy). Pyrolytic carbon is a material that has high strength, fatigue and wear resistance, and exceptional biocompatibility including thromboresistance. The sewing cuff, which anchors the valve into the native orifice, is composed of expanded polytetrafluoroethylene (ePTFE), Dacron or other fabric to allow suturing and subsequently tissue integration into the host tissue. The opening and closing of the valve is purely a passive phenomenon, with the moving parts [occluder or disk(s)] responding to changes in pressure and blood flow within the chambers of the heart and great vessels. Patients receiving mechanical valves must be treated with lifelong anticoagulation to reduce the risk of thrombosis and thromboembolic events.

Tissue valves are anatomically more similar to natural valves than are mechanical prostheses. Most tissue valves are composed of three cusps of tissue derived from animals—most frequently either porcine (pig) aortic valve or bovine (cow) pericardium—treated with glutaraldehyde. This fixation preserves the tissues, kills the cells within the valve, and decreases the immunological reactivity of the tissue, so that no immunosuppression is required for these xenografts as is required for kidney or heart transplants. However, since these valves no longer contain viable cells, the cusps themselves cannot remodel or respond to injury as does normal tissue. These cusps are mounted on a metal or plastic stent with three posts (or struts) to stimulate the geometry of a native valve.

Peripheral Stents and Stent Grafts

Peripheral vascular disease results mainly from narrowing of the aorta and its branches secondary to atherosclerosis, the same accumulation of plaque within the arterial wall were formed newly. As the degree of stenosis increases, blood flow to the distal tissues is impeded, causing ischemia in the tissues served by the diseased artery. The treatment for many patients with peripheral vascular disease involves using a vascular graft to perform a bypass around the area of blockage to restore ample blood flow. Surgical procedures such as open abdominal aortic aneurysm repair and aorta-femoral bypass grafting can have significant associated complications in certain patient populations. A minimally invasive approach such as that afforded by percutaneous transluminal angioplasty (PTA) with or without stenting is appropriate in these settings.

Stents and stent grafts can be employed in the peripheral circulation to increase lumen size in a similar fashion to their use in the coronary circulation. Stents for treatment of peripheral vascular disease are generally constructed of stainless steel or nitinol and may be coated with compounds such as ePTFE (expanded polytetrafluoroethylene). Stent grafts are composed of a metallic frame covered by a fabric tube and combine the features of stents and vascular grafts; they can be deployed endovascularly. Stent grafts are used to treat aortic aneurysms, where the aortic wall has been weakened and threatens to rupture, as well as stenosis of other arterial sites.

The graft portion, usually composed of polyester or ePTFE, can sit on either the luminal or abluminal aspect of the metallic stent and is intended to provide a mechanical barrier to prevent intravascular pressure from being transmitted to the weakened wall of the aneurysm. These stents and stent grafts are deployed in a similar manner to those in the coronary circulation, either as self-expanding units or over an inflatable balloon. The stent used for a given application is selected by diameter, length and geometry of the lesion and location of side branches or branch points. Stents and stent grafts have been especially successful in treating subtotally occluded short (5–10 cm) segments of the iliac artery that can cause significant chronic lower extremity ischemia, and in the treatment of stenosis of the renal arteries and the smaller arteries of the lower extremity.

Orthopaedic Applications

Orthopaedic biomaterials are enormously successful in restoring mobility and quality of life to millions of individuals each year. Orthopaedic implants include reconstructive implants, fracture management products, spinal products, rehabilitation products, arthroscopy products, electrical stimulation products and casting products. These products are generally used for either fracture fixation or joint replacement. More specific orthopaedic applications within these two categories are listed below.

Fracture fixation devices

1. Spinal fixation devices
2. Fracture plates
3. Wires, pins and screws
4. Intramedullary devices
5. Artificial ligaments

Joint replacement

1. Hip arthroplasty
2. Knee arthroplasty
3. Ankle arthroplasty
4. Shoulder arthroplasty
5. Elbow arthroplasty
6. Wrist arthroplasty
7. Finger arthroplasty

Orthopaedic Biomaterials

Orthopaedic biomaterials are generally limited to those materials that withstand cyclic load-bearing applications. Although metals, polymers, and ceramics are used in orthopaedics, it is metals that have over the years, uniquely provided the appropriate material properties such as high strength, ductility, fracture toughness, hardness, corrosion resistance, formability and biocompatibility that are necessary for most load-bearing roles required in fracture fixation and total joint arthroplasty (TJA). The use of orthopaedic biomaterials generally falls into three surgical specialty categories: upper extremity, spine or lower extremity. Each specialty is typically divided into three general categories: paediatric, trauma and reconstruction. Despite these numerous specialties and the hundreds of orthopaedic applications, there are only a few orthopaedic metals, ceramics and polymers that dominate. Knowing the general properties, uses and limitations of the "primary" orthopaedic biomaterials is requisite to understanding what is required to improve the performance of current implant materials and why only a few dominate the industry. A summary of seven more-prevalent orthopedic biomaterials and their primary use(s) are listed in Table 5.1.

Table 5.1 Most Common Orthopaedic Biomaterials

Material	Primary use(s)
Metals	
Ti alloy (Ti-6% Al-4% V)	Plates, screws, TJA components (nonbearing surface)
Co-Cr-Mo alloy	TJA components
Stainless steel	TJA components, screws, plates, cabling
Polymers	
Poly(methyl methacrylate) (PMMA)	Bone cement
Ultrahigh-molecular-weight polyethylene (UHMWPE)	Low-friction inserts for bearing surfaces TJA
Ceramics	
Alumina (Al_2O_3)	Bearing-surface TJA components
Zirconia (ZrO_2)	Bearing-surface TJA components

Orthopaedic Biomaterials Design

The developer of new biomaterials for orthopaedic purposes faces the same duality of concerns present in all other implant use: 1) the material must not adversely affect its biological environment, and 2) in return the material must not be adversely affected by the surrounding host tissues and fluids. And it must do this in a way that exceeds that of present materials used for the same application, if any exist. In order for new "improved" orthopaedic biomaterials there should be some understanding of the interrelationship between the structure and properties of the "form-function" relationship in calcified tissues will help provide an insight into factors determining implant design as well as deciding which are the materials of choice to meet a specific orthopaedic need.

Polymers are most commonly used in orthopaedics as articulating bearing surfaces of joint replacements and as an interpositional cementing material between the implant surface and bone. Polymers used as articulating surfaces must have low coefficients of friction and low wear rates when in articulating contact with the opposing surface, which is usually made of metal. Polymers used for fixation as a structural interface between the implant component and bone tissue require appropriate mechanical properties of a polymer, which can be moulded into shape and cured *in vivo.* The first type to be used, PMMA [Poly(methyl methacrylate)], is a grouting material to fix both the stem of the femoral component and the acetabular component in place, and thus distribute the loads more uniformly from the implants to the bone. Since high interfacial stresses result from the accommodation of a high-modulus prosthesis within the much lower modulus bone, the use of lower modulus interpositional material has been a goal of alternatives seeking to improve upon PMMA fixation. Thus, polymers such as polysulphone have been tried as porous coatings on the implant's metallic core to permit mechanical interlocking through bone and/or soft tissue ingrowth into the pores. However to date PMMA remains the method of choice for orthopaedic surgeons. This requires that polymers have surfaces that resist creep under the stresses found in clinical situations and have enough yield strength to minimize plastic deformation. The important mechanical properties of orthopaedic polymers are yield stress, creep resistance and wear rate. These factors are controlled by such parameters as molecular chain structure, molecular weight and degree of branching or chain linearity.

Orthopaedic Biomaterials: Clinical Concerns

Current implant designs represent the cumulative efforts of scientists, engineers and physicians and generally last 10–20 years with an approximate revision rate of 7% after 10 years of service. The benefits provided to patients by orthopaedic implants in terms of pain, mobility and quality of life are immeasurable. Therefore the following sections, which focus on the problems associated with implants, seek to provide the student of biomaterials a foundation for understanding the relevant issues for orthopaedic research and is not intended to serve as an indictment of orthopaedic materials.

Despite their success over the long term (≥ 7 years), orthoapaedic biomaterials have been associated with adverse local and remote tissue responses. It is generally the degradation products of orthopedic biomaterials (generally by wear and electrochemical corrosion) that mediate these adverse effects. This debris may be present as particulate wear, colloidal nanometre-sized complexes (specifically or nonspecifically bound by protein), free metallic ions or inorganic metal salts/ oxides or in an organic storage form. Clinical aspects of biocompatibility regarding polymer and metal release from orthopaedic prosthetic devices have taken on an increasing sense of urgency because of the escalating rates of people receiving implants and the recognition of extensive implant debris within local and remote tissues. Particulate debris have enormous specific surface areas available for interaction with the surroundings and chronic elevations in serum metal content. Clinical issues associated with biomaterial degradation can be broken down into some basic questions on degraded materials.

Orthopaedic Biomaterials Wear

The generation of wear debris, and the subsequent tissue reaction to such debris, is central to the longevity of total joint replacements. In fact, particulate debris are currently extolled as a primary factor affecting the long-term performance of joint replacement prostheses, and the primary source of orthopaedic biomaterial degradation. Particulate debris generated by wear, fretting or fragmentation includes the formation of an inflammatory reaction, which at a certain point promotes a foreign-body granulation tissue response that has the ability to invade the bone-implant interface. This commonly results in progressive, local bone loss that threatens the fixation of both cemented and cementless devices alike.

Mechanisms of wear debris generation Wear involves the loss of material in particulate form as a consequence of relative motion between two surfaces. Two materials placed together under load will only contact over a small area of the higher peaks or asperities. Electrorepulsive and atomic binding interactions occur at the individual contacts and when the two surfaces slide relative to one another, these interactions are disrupted. This results in the release of materials in the form of particles (wear debris). The particles may be lost from the system, transferred to the counterface or remain between the sliding surfaces. The following three processes can cause wear:

1. abrasion, by which a harder surface "plows" grooves in the softer material,
2. adhesion, by which a softer material is smeared onto a harder countersurface, forming a transfer film, and
3. fatigue, by which altering episodes of loading and unloading result in the formation of subsurface cracks which propagate to form particles that are shed from the surface.

Orthopaedic Biomaterial Corrosion

Electrochemical corrosion occurs to some extent on all metallic surfaces including implants. This is undesirable for two primary reasons: 1) the degradative process may reduce structural integrity of the implant and 2) the release of products of degradation is potentially toxic to the host. Metallic biomaterial degradation may result from electrochemical dissolution phenomena or wear, but most commonly occurs through a synergistic combination of the two. Electrochemical processes include generalized corrosion uniformly affecting an entire surface and localized corrosion affecting either areas of the device relatively shielded from the environment or seemingly random sites on the surface.

Corrosion of orthopaedic biomaterials is a multifactorial phenomenon and is dependent on five primary factors: 1) geometric variables, 2) metallurgical variables, 3) mechanical variables (stress and relative motion), 4) solution variables (pH, solution proteins, enzymes), and 5) the mechanical loading (degree of movement, contact forces). Current investigational efforts to minimize the corrosion of the orthopaedic biomaterials deal directly with the complex interactions of these factors. There are two essential features associated with how and why a metal corrodes. The first has to do with thermodynamic driving forces, which cause corrosion reactions. The second factor governing the corrosion process of metallic biomaterials is the kinetic barrier to corrosion.

CONCEPT CHECK

1. Why is sustained or controlled drug delivery important in the field of medicine?
2. How will controlled drug delivery help for prolonged circulation?
3. How and why is a nano-sized delivery vehicle suitable for targeted drug delivery?
4. List any biomaterials as examples for orthopedic applications.

MIND TWISTER

1. In which range of size will our immunological system get alert?
2. What will be the fate of nanoparticles after drug delivery?
3. Do degraded nanoparticles have any harmful effect on our body system?

REFERENCES

Cooke, F.W. (1996). Bulk properties of materials. In: *Biomaterials Science: Introduction to Materials in Medicine.* Ratner, B.D., Hoffman, A.S., Schoen, F.J. and Lemons, J.E. (eds.). Academic Press, London.

Crespo, L., Sanclimens, G., Montaner, B., Perez-Tomas, R., Royo, M., Pons, M., Albericio, F. and Giralt, E. (2002). Peptide dendrimers based on polyproline helices. *J. Am. Chem. Soc.* 124: 8876–8883.

Fann, J.I. and Burdon, T.A. (2001). Are the indications for tissue valves different in 2001 and how do we communicate these changes to our cardiology colleagues? *Curr. Opin. Cardiol.* 16: 126–135.

Faries, P., Morrissey, N.J., Teodorescu, V., Gravereaux, E.C., Burks, Jr., J.A., Carroccio, A., Kent, K.C., Hollier, L.H. and Marin, M.L. (2002). Recent advances in peripheral angioplasty and stenting. *Angiology.* 53: 617–626.

Fattori, R. and Piva, T. (2003). Drug-eluting stents in vascular intervention. *Lancet.* 361: 247–249.

Harding, K.G., Morris, H.L. and Patel, G.K. (2002). Science, medicine and the future: healing chronic wounds. *Br. Med. J.* 324: 160–163.

Kim, G.M., Bae, Y.H. and Jo, W.H. (2005). pH-induced micelle formation of poly(histidine-co-phenylalanine)-block-poly(ethylene glycol) in aqueous media. *Macromol. Biosci.* 5: 1118–1124.

Newkome, G.R., Yao, Z.Q., Baker, G.R. and Gupta, V.K. (1985). Cascade molecules: A new approach to micelles, A[27]-arborol. *J. Org. Chem.* 50: 2003–2006.

Ramkissoon-Ganorkar, C., Liu, F., Baudys, M. and Kim, S.W. (1999). Modulating insulin-release profile from pH/thermosensitive polymeric beads through polymer molecular weight. *J. Control. Release.* 59: 287–298.

Ratner, B.D. (2001). Replacing and renewing: synthetic materials, biomimetics, and tissue engineering in implant dentistry. *J Dent. Educ.* 65: 1340–1347.

Tomalia, D.A., Baker, H., Dewald, J.R., Hall, M., Kallos, G., Martin, S., Roeck, J., Ryder, J. and Smith, P. (1985). A new class of polymers: Starburst-dendritic macromolecules. *Polym. J.* 17: 117–132.

Tomalia, D.A., Naylor, A.M. and Goddard III, W.A. (1990). Starburst dendrimers: Molecular-level control of size, shape, surface chemistry, topology, and flexibility from atoms to macroscopic matter. *Angew. Chem. Int. Edn.* 29: 138–175.

Webb, J.C.J. and Tricker, J. (2000). Bone biology: a review of fracture healing. *Curr. Orthop.* 14: 457–463.

Wunderbaldinger, P., Josephson, L. and Weissleder, R. (2002). Tat peptide directs enhanced clearance and hepatic permeability of magnetic nanoparticles. *Bioconjug. Chem.* 13: 264–268.

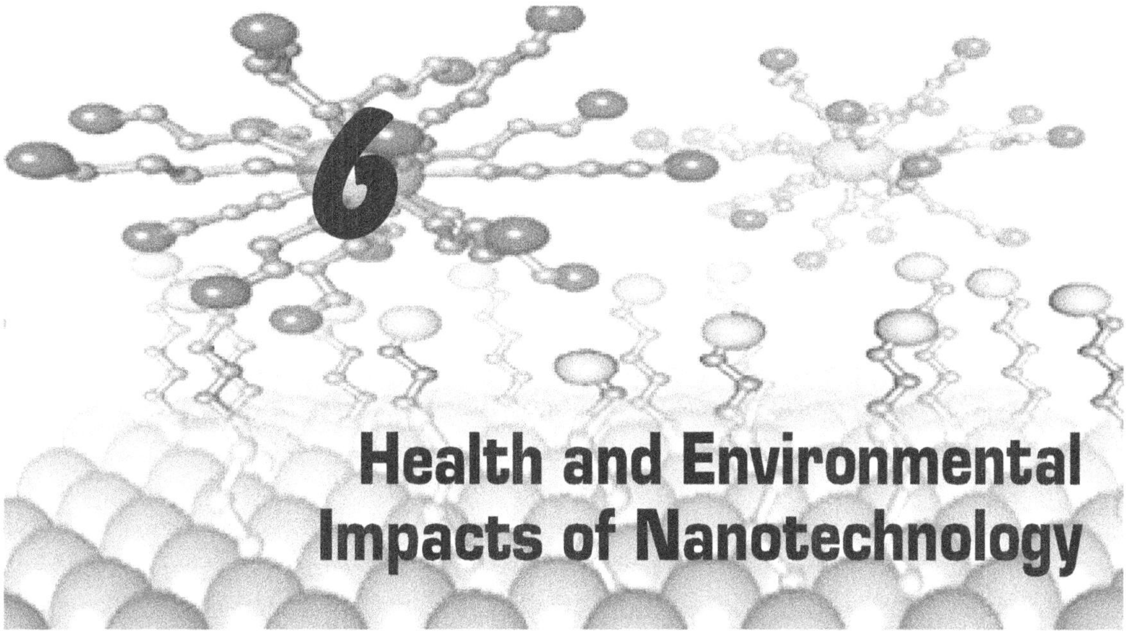

6 Health and Environmental Impacts of Nanotechnology

INTRODUCTION

Nanotechnology has variously been described as a transformative technology, an enabling technology, and the next technological revolution. However, enthusiasm over the rate of progress is being tempered increasingly by concerns over possible downsides of the technology, including unforeseen or poorly managed risk to human health. Real and perceived adverse consequences in areas such as asbestos, nuclear power, and genetically modified organisms have engendered increasing skepticism over the ability of scientists, industry, and governments to ensure the safety of new technologies. As nanotechnology moves toward widespread commercialization, not only is the debate over preventing adverse consequences occurring at an unusually early stage in the development cycle, it is also expanding beyond traditional knowledge-based risk management to incorporate public perception, trust, and acceptance.

Within this context, the long-term success of "nanotechnologies" (referring to the many specific applications and implementations of nanotechnology) will depend on rational, informed, and transparent dialogue aimed at understanding and minimizing the potential adverse implications to human health and the environment. A central question that has been raised by everyone is "how safe is nanotechnology?" However, such a general and unbound question is unlikely to yield useful information on the safety of specific nanotechnologies without further contextual information. Rather, appropriate contexts need to be defined and boundary conditions set if information on the safety of specific nanotechnologies is to be developed.

ENGINEERED NANOMATERIALS OF RELEVANCE TO HUMAN HEALTH

Nanotechnologies will likely be so diverse as to defy generic classification when it comes to evaluating potential health impact. It is therefore important to be able to define criteria that distinguish between technologies and products more or less likely to present a health risk, if we are to avoid inappropriate and possibly deleterious sweeping conclusions regarding potential impact. For example, complementary metal-oxide-semiconductor devices with sub-100 nm features, or high-resolution electron microscopes, will present a fundamentally different potential risk to human health than products containing unbound nanostructured particles, such as nanophase zinc oxide-based sunscreens. It is anticipated that nanotechnology standards being developed by organizations such as the International Standards Organization (ISO) and ASTM International will arrive at appropriate criteria in due course. In the meantime, a number of published works have hinted at or proposed working criteria. The 2004 report on nanotechnology from the Royal Society and Royal Academy of Engineering highlighted nanotechnologies associated with unbound sub-100 nm diameter particles as being of particular interest to human health and support this emphasis on sub-100 nm diameter particles in a discussion on the emerging field of nanotoxicology. However, it is clear from published toxicity studies that particle size alone is not a good criteria for differentiating between more or less hazardous materials and technologies. For instance, inhalation studies using rodents have demonstrated that 20 nm diameter TiO_2 particles have a greater impact on the animals' lungs than pigment-grade particles with the same composition, even though both particle sizes were administered as micrometre-diameter agglomerates. So it is perhaps more appropriate to address the potential health impact of nanostructured particles—those having sub-100 nm scale structures—than nanometre diameter particles. By exploring this idea further, noting that the scale-dependent properties of nanomaterials are not necessarily associated with particle diameter, but with material structure. As an example, they use open agglomerates of single-walled carbon nanotubes (SWNTs), which may be micrometres in diameter, but exhibit structure at the nanoscale that is likely to influence their behaviour.

Within the context of inhalation exposure, nanomaterials must meet the following criteria in order to present a unique potential risk to human health:

1. The material must be able to interact with the body in such a way that its nanostructure is biologically available;
2. The material should have the potential to elicit a biological response that is associated with its nanostructure.

Although these two criteria relate to inhalation exposure, they are sufficiently broad to encompass all potential routes of exposure, and provide a useful working framework for distinguishing between materials and products that are less likely to present a health risk and those that are more likely to have some potential for adversely affecting health. When these criteria are linked to potential exposure to the skin, respiratory system, and gastrointestinal (GI) system, categories of materials and sources begin to emerge that may present a greater risk under some circumstances. These include unbound nanometre-diameter particles (in powders, aerosols, and liquid suspensions); agglomerates and aggregates of nanometre-diameter particles, where nanostructure-based functionality is retained; aerosolized liquid suspensions of nanomaterials; and the attrition (or comminution) of nanomaterial composites through various mechanisms.

ENGINEERED NANOMATERIALS IN THE BODY

While quantitative risk analysis considers many factors, the potential for a material to cause harm (hazard potential), and the amount of material able to reach target organs within the body (exposure potential) are critical to understanding potential health impact. Paracelsus (1493–1541), widely regarded as the father of modern toxicology, is credited with the statement that "all things are poison and not without poison; only the dose makes a thing not a poison". As true now when dealing with emerging technologies as it was 500 years ago, his statement emphasizes the need to understand both how harmful a substance is, and how much of it can get into the body (and to specific organs), if risk is to be understood and managed.

ROUTES OF ENTRY

Three routes of entry into the body are likely to be of primary significance for engineered nanomaterials—inhalation, ingestion, and dermal penetration.

Two additional routes become important when considering nanotechnology-based medical devices and drugs–injection and release from implants. Focusing on nonmedical exposure, the literature on impact associated with inhalation exposure vastly outweighs the alternative exposure routes, reflecting a current research emphasis on the health impact of airborne nanostructured materials. Whether this represents relative risk, rather than the current interests of the research community, is unclear. Certainly, the health impacts of inhaling airborne particles have long been recognized: associations between exposure to 'very fine particles' and lung disease were recognized by Ramazzini in the 17th century, and documented links between aerosol exposure and ill health date back to the 4th century.

Gastrointestinal Tract

Particles deposited in the respiratory system that are cleared via the mucociliary escalator may be swallowed, leading to exposure to the GI tract. Additional ingestion routes include the use of nanostructured materials in food, water, and drugs. Relatively few studies have investigated nanostructured materials in the GI tract, and most have shown them to pass through and be eliminated rapidly. However, compared with inhalation and skin exposure routes, there does not appear to be much research currently focused on this potential route of entry.

Skin

There has been a greater focus on the skin as a potential route of entry in recent years. The inclusion of nanoscale particles in sunscreens and cosmetics has raised concerns over possible dermal penetration of material, leading to ill health. For example, nanoscale particles of materials such as TiO_2 and zinc oxide are being used as effective ultraviolet (UV) blocking agents in sunscreens, and nanoscale liposomes are currently used as delivery vehicles in skincare products. Dermal exposure and penetration are also potential issues when handling engineered nanomaterials. Whether engineered nanomaterials in contact with the skin represent a significant risk to health depends on their ability to penetrate through the outer protective layers and reach the epidermis or dermis, and the subsequent impact they may have on the body. Latex particles smaller than 1 μ m penetrate the outer layers of a skin sample during constant flexing. Other studies indicate that healthy, intact skin presents a good barrier against nanostructured particles. However, there are indications that hair follicles could act as a repository of nanometre-diameter particles, and that the chemistry of carrier liquids may affect penetration potential.

Recently, performed studies have shown that nanoscale quantum dots with different sizes, shapes, and coatings penetrate through the outer layers of pig skin samples in a flow cell, and enter the epidermal and dermal layers within 24 hours. The smallest particles (only 4.6 nm in diameter) showed localization in the epidermis and dermis within 8 hours, irrespective of the coating material used (polyethylene glycol, carboxylic acid, or polyethylene glycol-amine). Larger nonspherical particles (12 nm by 6 nm ellipsoids) showed a penetration rate that depended on the coating, but particles with all three coatings were found in the epidermis and dermis after 24 hours. Even if nanoscale particles are able to penetrate through the outer layers of the skin, there is very little information on the hazard they might present. Research using subcutaneously introduced nanoscale particles suggests that they can be transported within the lymphatic system, raising questions about how they might influence immune responses, and there are some indications that neuronal uptake and transportation may occur. However, discussions on the mechanisms of interaction and possible health outcomes are still rather speculative.

Some concern has been expressed that the photogeneration of hydroxyl radicals by nanosized particles of materials like TiO_2 and zinc oxide may lead to oxidative damage in the skin, although the use of surface modification in such nanoparticles has been shown to suppress free radical generation. Ingestion, and possibly dermal penetration, are likely to become increasingly significant exposure routes as engineered nanomaterials are used in an ever-widening range of products. A recent survey of nanotechnology-based consumer products found that, out of over 200 manufacturer-identified "nano" consumer products currently available, over 30% are applied directly to the skin or eaten. In addition to products like these that are intentionally introduced to the body, little is known about the environmental accumulation of nanomaterials over product life cycles, how this might affect exposure profiles.

Lungs

Inhalation of airborne material is clearly a significant potential exposure route. Aerosol penetration into and deposition within the respiratory system has been studied and modelled extensively. Most airborne particles smaller than a few tens of micrometres in diameter can be inhaled. Once in the respiratory system, particles will deposit in different regions according to their shape, diameter, and density. Diffusion-based deposition mechanisms lead to relatively high particle deposition probability in the alveolar region of the lungs for particles smaller than approximately 300 nm (although particles smaller than 4 μm have a greater than 50% probability of penetrating to this region). Below approximately 30 nm, diffusion leads to high deposition probabilities throughout the respiratory system, including the upper airways. At smaller diameters, deposition in the upper airways and especially the nasal region begins to dominate deposition in the alveolar region.

Evaluations of health risk associated with aerosol exposure are generally based on the assumption that toxicity is associated with the mass and chemical composition of inhaled material. This mass-based approach has been very effective historically, leading to substantial reductions in respiratory disease with reduced exposures. However, recent research has challenged the robustness of this approach for inhaled low-solubility particles. In one study, rats exposed to less than 60 $\mu g/m^3$ of freshly generated 26 nm diameter polytetrafluoroethane (PTFE) particles died of haemorrhagic pulmonary inflammation in less than 30 minutes. PTFE is a chemically inert polymer, and mortality was observed at mass concentrations comparable to the daily $PM_{2.5}$ standard in the US, and one hundred times lower than occupational exposure limits for respirable "nuisance dusts."

Some researchers have further demonstrated a particle size dependence on pulmonary inflammatory response in rats using TiO_2 particles. Although smaller particles were shown to be more potent than larger ones on a mass concentration basis, different sized particles showed a similar response when dose was interpreted in terms of particle surface area. Similar studies have shown that response to low-solubility particles scales poorly with mass concentration, but closely with surface area concentration. The dose-response relationship appears to be similar for chemically inert, low-solubility materials, suggesting a mechanism associated with the physical nature of the particles. However, insoluble particles that are chemically active, such as crystalline quartz, remain markedly more toxic than other insoluble materials, even when normalized for surface area.

Particle shape is also a factor when addressing potential hazard. Exposure to anisotropic particles such as fibres (e.g. asbestos) has long been associated with increased risk of fibrosis, lung cancer, and mesothelioma. This raises additional concerns over the role of particle morphology when considering some complex nanostructured materials, including nanometre-diameter tubular and fibrous structures. SWNTs can elicit transitory inflammation in rats and lead to multifocal granulomas, when introduced to the lungs using intratracheal instillation. Some researchers have demonstrated that pharyngeal aspiration of SWNTs in mice leads to acute inflammation with early onset yet progressive fibrosis and granulomas. Fibrosis was associated with dense clumps of deposited SWNT material, but was also observed in distal regions of the mice lungs where dense clumps of SWNTs were not seen. Observations of fibrosis in the absence of clearly visible SWNTs led to the hypothesis that different SWNT agglomerate morphologies are responsible for the two distinct responses observed.

Although multiwalled carbon nanotubes (MWNTs) are already used in commercial products (predominantly encapsulated in composite materials), relatively few studies have investigated their potential toxicity. In a recently published *in vitro* study, MWNTs were found to penetrate into human epidermal keratinocyte cells and to elicit production of an inflammatory cytokine and were shown to be cytotoxic to human T-cells, with oxidized nanotubes being significantly more potent than pristine nanotubes.

TOXIC MECHANISMS

A mechanistic understanding of nanostructured material behaviour in the body and ill health is still some way off, although a number of research articles address possible mechanisms of interaction. Particles that enter the bloodstream may affect the blood vessel lining or function and promote blood clot formation, they may also be associated with cardiovascular effects linked to inhaling ambient ultrafine particles. A cardiovascular response initiated by lung inflammation has also been proposed that does not depend on particles entering the bloodstream. Although somewhat speculative, computer modeling has indicated that C60 molecules may bind to and deform nucleotides, if they are able to come into contact with DNA molecules. While subcellular exposure to free underivatized C60 molecules (which are hydrophobic) is unlikely, C60 can form stable nanometre-diameter colloidal particles in water, and may potentially penetrate cells in this form. Oxidative stress is considered to be an important mechanism, and certainly a diverse range of nanoscale materials have been shown to generate reactive oxygen species (ROS) in biological environments. Even so, there remains considerable uncertainty over the processes underlying ROS generation from particles, and the precise impact on organ-level, cellular, and subcellular systems.

IMPORTANCE OF MATERIAL CHARACTERIZATION

The dependence of engineered nanomaterials' behaviour on physical and chemical structure significantly increases the difficulty in developing a sound understanding of material toxicity. Without detailed material physicochemical characterization, toxicity studies become difficult to interpret, and inter-comparison of studies becomes near impossible. Factors such as agglomeration state, surface chemistry, material source, preparation method, and storage take on a significance that has often been overlooked, potentially leading to inappropriate conclusions being drawn. For example, the toxicity of a material such as a SWNT is likely to be affected significantly by production process, atomic structure, surface modification, purity, aggregate morphology, preparation method, and method of delivery. Without information at this level, comparing toxicity evaluations becomes highly qualitative. When evaluating engineered nanomaterials, it has been stated in the final report that, "It is essential that the physical and chemical characterization of nanoscale materials be much more complete than has been the case in the sparse toxicology literature appearing to date".

ENVIRONMENTAL IMPLICATIONS OF NANOPARTICLES

Nanotechnology's promise is that it will provide new means for pollution remediation and less toxic ways to manufacture goods. However, latent toxicity is the flipside: nanosize materials are interesting because their physical and chemical properties differ so radically from bulk amounts of chemically the same compounds. Some of these differences are useful and wanted, but others have the potential to be less desirable. In the way of specifics, nanotechnology is likely to enable more efficient and effective water filtration, options for cleaning up oil spills, various coatings (to protect against the environment) and, potentially, artificial photosynthesis. But these applications carry risks with them as well. For example, one study has linked buckyballs (i.e., synthetic carbon molecules of a specific orientation) to brain damage in fish. Also, studies have indicated that carbon nanotubes can lead to toxic effects in mice. Certainly there are issues worth discussing in these studies, in particular, the delivery mechanism in the mice study is unlikely to be naturally occurring, but they at least highlight some of the potential impacts that nanotechnology could have on the biological world. And, of course, the environmental implications of nanotechnology could (directly) affect humans as well. Particularly at risk could be those who work in factories where production might lead to the liberation of nanoparticles into the air; worker safety is surely a legitimate ethical concern.

In addition to toxicity, there are other ethical issues that pertain to nanotechnology and the environment. For example, as water purification becomes more efficient and effective, it might be the case that we then incur duties to apply these technologies (as against inefficient and ineffective previous generations thereof). Furthermore, the availability of environmentally positive nanotechnologies may change the moral status the developed world bears to the less developed (and, in particular, the environmentally compromised) world. Most basically, then, there will be questions about whether we can use these technologies (given toxicity and other risks), whether we have to use these technologies (given obligations of environmental stewardship), and whether we have to share them (given obligations to international distributive justice).

There are many types of nanomaterials (NMs) and the scientific community is making observations on nanoparticle (NP) ecotoxicity to inform the wider debate about the risks and benefits of these materials. As we already discussed about this, that NMs have been defined as material with at least one dimension between 1 and 100 nm. However, this is a somewhat arbitrary definition, and for ecotoxicology, we should also consider NMs with a distribution of particle sizes around the nanoscale that may include some primary particles larger than 100 nm; or larger aggregates of NPs of a few hundred nanometres. In mammalian respiratory toxicology, sizes of particulate matter (PM) have been traditionally defined as coarse particles (diameter between 10 and 2.5 μm), fine particles (2.5 μm or less), or ultrafine particles (<0.1 μm). Nanoparticles can therefore also be regarded as ultrafine particles, and emerging ecotoxicological data might be compared with some of the ultrafine particle exposures performed on rodents. Manufactured NMs are designed to achieve particular physico-chemical properties that relate to the product application. The materials can be carbon-based such as carbon spheres, carbon nanotubes, metal-based NPs, composite NMs or multilayer NPs (e.g. platinum core silica shell, Pt@SiO_2), or NPs with an external coating or capping agent (e.g. functionalized zinc oxide NPs). There are currently many types of nanoproducts on the market, and many potential benefits of these new materials. The products and applications include electronics, optics, textiles, medical devices, cosmetics, food packaging, water treatment technology, fuel cells, catalysts, biosensors and agents for environmental remediation. There is now a wider debate about the risks and benefits of the many manufactured NMs and consumer products, and this includes consideration of risks to the environment. The ecotoxicology community is only at the beginning of understanding the potential risks to wildlife associated with manufactured NMs. We recognize that these materials may have unusual physico-chemical properties, or behaviours in water (e.g. colloid chemistry), that are less familiar to ecotoxicologists compared to the properties of other pollutants like metals or pesticides. The application of these materials (i.e., nanotechnology) is also relatively new.

Clearly, the scientific debate on the environmental safety of NMs needs to adopt a multidisciplinary approach involving physicists, chemists, material scientists, biologists, toxicologists, risk assessors, regulators and policy makers. In order to have such a debate, the Society for Environmental Toxicology and Chemistry-UK branch (SETAC-UK) recently organized a meeting called the "2nd International Conference on the Environmental Effects of Nanoparticles and Nanomaterials" to bring this diverse group of professionals together. This meeting, aimed to highlight the key themes, complexities, controversies and knowledge gaps on the ecotoxicology and environmental chemistry of manufactured NMs. In addition, to recognize that nature has been producing natural NPs and NMs for millions of years, and so we also aim to reflect on our knowledge of these natural materials and whether manufactured NMs represent a new or additional risk to the environment.

The physico-chemistry above highlights a number of key issues for the ecotoxicologist. These include:

i. The fate and behaviour of most manufactured NPs are likely to be different in sea water compared to fresh water, and therefore ecotoxicity may also be different.

ii. Ecotoxicity in fresh water may be particularly influenced by the presence of organic matter, changes in pH, and the presence of cations such as Ca^{2+}.

iii. Environmental exposure will not be homogeneous. Adsorption and aggregation phenomena may promote local high concentrations of manufactured NPs in sediments, on biofilms, or in microsurface layers (e.g. ocean microsurface layer).

iv. Adsorption phenomena may drive surface acting toxicity to organisms, and this could occur without appreciable bioaccumulation within the organisms.

v. Uptake into the organism will depend on the aggregation chemistry on the exterior surfaces of the organism, and the behaviour of manufactured NPs in body fluids such as plasma.

vi. Toxicity may be a function of particle size and shape, and that there may be inherent differences in the toxicity of NPs compared to micron scale particles.

In addition, it is also worth considering how the above chemistry will impact on the behavior of NPs in polluted environments, and on whether NPs are likely to interact with other (non-nano) pollutants in the environment. In theory, the presence of surfactants in polluted environments might stabilize manufactured NPs in the aqueous phase (as they would for natural NPs above). Alternatively, some types of organic matter and particulate materials in effluents might cause NP aggregation. It is also possible that manufactured NPs could adsorb organic chemicals to the outer surface of the particle. This might reduce the bioavailability of the chemical, or alternatively, the NP may act as a delivery vehicle for the organic chemical.

TOXICOLOGICAL HEALTH EFFECTS CAUSED BY NANOPARTICLES

Toxicological experiments can be categorized into *in vivo* and *in vitro* studies. *In vivo* experiments investigate effects in living organisms such as experimental animals or healthy subjects and patients in clinical studies, whereas *in vitro* experiments are conducted in organs, tissues, cells or biomolecules isolated from the living organism. In general, toxicology studies on air pollutants are shorter-term experimental approaches that tend, for ethical reasons, not to study people but experimental animals. They often analyze the early events rather than waiting for final disease. In addition, in those studies high doses are required to detect significant effects; and they may use cells and biochemical systems rather than whole animals. To compensate those shortcomings, models of susceptibility are an important experimental approach in which the biological system is predisposed prior to the treatment with air pollutants. The predisposition of the system acts already like a change of the homeostatic balance and may respond more vigorously than the normal system. If in addition, the predisposition reflects or resembles human diseases or human predisposition; such a model may be a powerful analytical tool, providing insights on how air pollutants may change responses of the susceptible system. While human clinical studies provide closer insight into nanoparticle-related modes of action, they are limited because of ethical reasons to yield firm dose–response relationships. The latter, however, can be provided by toxicological studies on susceptible animal models as well as cell systems. Hence, animal studies are supposed to provide biological explanations and plausibility and create hypotheses which may then be proven in human clinical studies.

Pulmonary Inflammation Induced by Ultrafine Particles

Many of the effects that occur rapidly after ultrafine particles deposit on the respiratory epithelium are not fully understood. Cells in contact with ultrafine particles like macrophages, epithelial cells and neutrophilic granulocytes are activated and may synthesize compounds referred to as reactive oxygen or nitrogen species (including free radicals, hydrogen peroxide, and superoxide) that try to inactivate and eliminate the invading foreign material. Within hours, cytokines and chemokines are synthesized and secreted into the affected area. These molecules are mediators that interact with specific receptors on the surfaces of many cell types and result in activating cells in the surrounding environment as well as in the blood and other tissues. As a result, cells leave the bloodstream and enter the interstitial spaces, where they can attack the foreign material. Consequently, particle-induced cell activation events in the airways frequently result in an inflammatory response. This response includes both, the activation of cells of the epithelium (including the production of the "pro-inflammatory" and reactive oxygen molecules described above) and the activation and migration of cells (particularly, neutrophilic granulocytes and eosinophilic granulocytes in case of a specific immunological response) from the blood into the airways. While the inflammatory response may occur after interaction of both fine and ultrafine particles with respiratory tissues, the enhanced surface area of ultrafine and nanoparticles compared to fine particles (as the acting interface) suggests a more prominent reaction.

Airway nerve cells may also contribute to inflammation in the airways by synthesizing neurotransmitters. In this neurogenic inflammation, the neurotransmitters can affect many types of white blood cells in the lung, as well as epithelial and smooth muscle cells. Inflammatory cytokines synthesized by white blood cells may also affect the nerve cells. The inflammatory response may damage the epithelial cell layer at the surface of the tissue and other cells in the airway (such as macrophages), which results in the loss of integrity of the tissue's defences. One potential consequence may be increased exposure to and reduced capacity to defend against microorganisms. Thus, particle deposition on the respiratory epithelium can trigger a cascade of events in many different cells, potentially resulting in changes in tissues and organs at sites progressively further away from the initial stimulus. These defence mechanisms are normal responses in healthy individuals, but they may lead to deleterious changes in the host. Such changes may be rapid and temporary and may resolve quickly; but depending on the level and pattern of exposure and the agent to which the host is exposed to, the changes may last longer. It is not clear whether or how such modulations are relevant to the development of particle-induced adverse health effects at low levels of exposure. Yet, these changes are thought to have a greater impact on individuals whose respiratory, cardiac, or vascular tissues have been previously altered or damaged. One possible consequence of damage to the airways is that the individual may become more susceptible to respiratory infections if exposed to viruses or bacteria. A second possible consequence is that it may further decrease respiratory function in a person, whose airways are already damaged by diseases, such as bronchitis or asthma. As a result symptoms of asthma, for example, may exacerbate.

Systemic Inflammation and Other Responses

Recent studies have suggested that exposure to particles results in systemic inflammatory effects within hours after exposure. Pathways are discussed to be either via direct particle translocation into circulation or via mediators released in the respiratory tract. Particularly, the former pathway supports the concept of enhanced cardiovascular responses after ultrafine or nanoparticles exposure due to the higher likelihood of ultrafine particle translocation into circulation when compared with larger particles. Recently a panel of cardiologists has reviewed the existing literature and compiled a statement in which possible pathways of interference of particulate matter have been studied and they link particle exposure with cardiovascular disease via four major routes: pulmonary inflammation, pulmonary reflexes and systemic translocation to circulation and to the heart. It is currently not clear whether the systemic response is a consequence of an inflammatory response in the respiratory tract, because some studies on systemic inflammation have detected little or no inflammatory lung response after exposure to PM. As described above, there are studies indicating that either particles per se (ultrafine and nanoparticles in particular) or components, that may detach or dissolve from particles, may move rapidly into the circulation triggering either oxidative stress or pro-thrombotic or acute-phase or other responses of the cardiovascular system. Therefore, a direct systemic inflammatory response is possible via particle translocation into the circulation. This pathway of systemic inflammation is thought to be capable of triggering a cascade of responses leading eventually to atherosclerotic plaque rupture and/or thrombosis as precursors of myocardial infarction. The other two pathways of pulmonary reflexes and direct particle action on the heart are thought to interfere with heart rate functions leading to arrhythmia. Both myocardial infarctions and arrhythmia are considered to be severe diseases with a substantial mortality rate.

RELEVANT PARAMETERS IN NANOPARTICLE TOXICOLOGY

Biologically it appears not meaningful to presume that low-solubility ultrafine or nanoparticles are hazards per se, since adverse reactions result from the interaction between the ultrafine particle and biological tissues. Currently certain parameters of ultrafine particles are considered to trigger or mediate a cascade of reactions starting with the formation of free radicals, which lead to oxidative stress in extracellular matrix and cells with the subsequent onset of pro-inflammatory processes.

Number Concentration and Surface Area

Considering health effects initiated by the exposure to ultrafine and nanoparticles requires a change of the paradigm that effects are correlated with the mass of the noxae accumulated during exposure. In ambient air the mass concentration of ultrafine particles is usually less than 10% of the mass concentration of $PM_{2.5}$. However, the number concentration of ultrafine particles dominates the number concentration of fine particles by >90%. Therefore, surface area and number concentration appear to be the more reasonable metrics of ultrafine and nanoparticles exposure than mass concentration. Furthermore, exposure metrics may be inadequate since it may be the number of deposited particles per unit surface area or dose to a specific cell (e.g. alveolar macrophage) that determines response for specific regions. Therefore, the use of a metric depends on specific questions posed, requiring specifically defined metrics. Ultrafine TiO_2 with an average particle size of 20 nm and pigment grade (fine) TiO_2 with an average particle size of about 250 nm were used. Doses ranging from 30 to 2000 μg of

TiO_2 were intratracheally instilled into rats and mice. When the deposited TiO_2 dose was expressed as particle surface area, there was a unique relationship to the inflammatory responses of these two different sizes of TiO_2 particles. The importance of particle surface area for eliciting inflammatory responses in the lung has been confirmed. This concept of particle surface area as the appropriate dose metric has been recognized as an important principle in particulate matter toxicology.

Particle Shape (Fibres and Nanotubes)

Newer materials or those that are under development, such as synthetic organic fibres and carbon nanotubes, may have different toxicology paradigms. The existing paradigm for silicate fibres suggests that respirable fibre types vary in their ability to cause lung disease and that this can be understood on the basis of the length of the fibres and their biopersistence in the lungs. Because fibres are regulated on a fibre number basis and the hazard is understood on the basis of the number of long fibres, in fibre testing the dose should always be expressed as fibre number, not mass and the length and diameter distribution need to be known. Carbon nanotubes are long thin structures which can exhibit diameters of a few nanometres, while the length can be up to many thousands of nanometres. These could have very unusual toxicological properties, in that they share shape characteristics of both fibres and nanoparticles; such limited toxicology as presently existing supports the contention that these may be harmful to the lungs. Thus, the physiological relevance of these findings needs to be ultimately determined by conducting inhalation toxicity studies.

Transition Metals

Since more than 10 years, transitional metals have been suspected and proved to cause health effects. Convincing evidence was provided by a combination of epidemiologic and subsequent toxicological studies. The epidemiological study showed reduced effects of several morbidity endpoints as well as mortality in the population of Utah Valley, Utah, USA, during a 1-year period when a steel mill had been shut down, resulting in considerable reduction of transitional metals in ambient fine particles, while symptoms and metal-containing air pollution were high in the year prior and after the closure of the steel mill. Dust sampled from each period was applied in a human clinical study as well as in a mechanistic study demonstrating pulmonary injury, neutrophilic inflammation and increased airway responsiveness of the metal-rich samples using a sensitive animal model. Hence, the clinical and the toxicological animal study provided a better understanding of the modes of action of the ambient particles which had been found to be associated with adverse health effects within the population of Utah Valley.

Similar combined studies are underway investigating air pollution of the city of Hettstedt in former Eastern Germany which has a history of several centuries of non-ferrous metal mining and smelting in comparison with the city of Zerbst serving as a control in a nearby agricultural area. High $PM_{2.5}$ levels were shown to be associated with significant decline of lung function and with significant increase of prevalence for bronchitis, for otitis media, for frequent colds, and for febrile infections in three surveys on children over a decade in the 1990s. At the same time high levels of transitional metals, zinc, lead, copper, and cadmium were associated with allergic responses. Dusts from Hettstedt and Zerbst were studied in an allergic mouse model showing increased allergic responses and increased allergic sensitization. At the same time the capacity of this dust was shown to form radical species and in addition clinical human studies demonstrated that the dust obtained from Hettstedt induced distinct airway inflammation in healthy subjects with a selective influx of monocytes and increased generation of oxidant radicals.

Both series of combinatory "epi-tox" studies provide comprehensive evidence for the association of adverse heath effects in susceptible population groups and modes of action of these ambient particles in clinical and animal model studies or in *in vitro* studies. It needs to be emphasized that $PM_{2.5}$ is the basic dose metric in these studies and not ultrafine particle parameters. Note that in the second series of epidemiological studies ultrafine particle number concentration was measured in addition and dominated the number concentration of the fine particle fraction by 90%. Therefore, additional research is required to demonstrate whether the ultrafine particle fraction plays a role in the observed effects and modes of action and whether they even may drive the effects because of the very peculiar properties of the ultrafine particle fraction.

More recently, investigators started to find evidence that amongst the large variety of organic compounds, particularly, in the particulate fraction of ambient air originating from combustion processes, there are biologically highly reactive compounds like redox cycling quinones—oxidized and nitrated polyaromatic hydrocarbons, which can catalyse release of reactive oxygen species (ROS), leading to the induction of oxidative stress and inflammation. PM from the Los Angeles basin as well as organic extracts obtained from DEP (Diesel Exhaust Particles) induce a stratified oxidative stress response leading to Heme-oxygenase-1 expression, followed by activation of Jun kinase and pro-inflammatory interleukin-8 production and culminated in cellular apoptosis in parallel with a sharp decline of antioxidant levels. These effects were more prominent in the fine particle fraction than the coarse, and they were positively correlated with higher contents of organic carbon and polyaromatic hydrocarbons. Admittedly, the ultrafine particle fraction was not analysed explicitly; however, the organic carbon load is highly associated with the ultrafine particle fraction since this fraction predominantly originates from combustion processes. Pro-inflammatory effects in the respiratory tract are related to the particle content of redox cycling chemicals and are involved in the adjuvant effect of DEP in atopic sensitization. Cytotoxicity in epithelial cells and macrophages is the result of mitochondrial damage, which manifests as ultramicroscopic changes in organelle morphology, a decrease in the mitochondrial membrane potential, superoxide production, and ATP depletion.

INTEGRATED CONCEPT OF RISK ASSESSMENT OF NANOPARTICLES

Nanoparticles are expected to be used in a wide range of new technologies. To highlight the importance of sustainable risk assessment, let us consider the use of nanoparticles in medicine: in this field nanoparticles are designed for therapeutic drug delivery and/or imaging techniques in diagnostics to be administered directly to the patient as multi-functional drug nanocarriers. This may require (a) targeting across several membranes while the drug is sufficiently bound to the nanocarrier, and (b) specific and controlled release of the drug at the target site, allowing for increased efficiency of the drug in target organs or cells, while side effects are minimized in sensitive but not targeted organs and tissues. This also implies non-toxic effects of the carrier-nanoparticles. Although these may be future visions, nanomedicines are likely to represent the most challenging nanoparticles in terms of their safe and sustained application, since patients predisposed by their disease are likely to be more susceptible to any such treatments responding eventually more sensitively than healthy subjects. Hence, the rapid development of a multitude of nanoparticle applications needs to be complemented by assessing possible implications assuring a safe and sustainable handling of those nanoparticles. The challenge of an integrated application development and implication assessment is pro-active collaboration at the earliest stage in order to optimize functionality of the nanoparticle and to minimize its side effects without losses in terms of costs and time because of one-sided mis-management or unfocussed or delayed initiation of risk assessment. While most other nanoparticles are not aimed for the specific administration in human subjects, possible exposure scenarios still need to be considered.

Hence an integrated conception to estimate the health hazards of newly generated nanoparticles during their development is proposed. For this risk analysis, the whole life-cycle of the newly developed nanoparticle has to be considered including its scientific or industrial generation, its storage and distribution, its anticipated application and possible abuse, and finally its disposal. That means potential human incorporation may vary at different stages of the life cycle of nanoparticles and different groups of the population may be exposed. During generation an occupational group of healthy adult workers may be exposed to nanoparticles, while its anticipated application may possibly lead to exposure of the whole population, including susceptible people, like infants, children, elderly and diseased individuals. The challenge is a reasonable estimate taking into account the widespread use of mass-produced nanoparticles versus those produced in small quantities.

As a result, possible incorporation pathways can be foreseen for the respective usage of a given nanoparticle, which will include organs of uptake, organs involved in the distribution and secondary target organs. In all these organs the nanoparticles or their metabolic products eventually accumulate in different ways. Within these organs, interaction of the nanoparticles will take place at the level of proteins of body fluids and on the cell surface and within cells; this will eventually be the beginning of a cascade of reactions mediating or initiating adverse health effects leading to disease, as outlined above. Therefore, only those proteins and cells need to be examined that interact with the nanoparticle. Nevertheless, these may be more than those within the organ of predominant uptake and anticipated target organs or tissues. To assess the interaction between nanoparticles and biological materials, a strategic concept is necessary that

- classifies nanoparticles according to their chemical compounds, physical structure and, particularly, those at their surface,
- takes into account the possible exposures in different phases of the nanoparticle's life cycle,

- estimates delivered doses to the various biological systems, and
- stratifies toxicological assessment from simple high-throughput screening methods towards more complex in vivo studies only when required and based on the previous finding at a lower level of analysis.

Initially simple acellular tests may evaluate free radical formation, oxidative stress, antigen–antibody reactions, etc. followed by genetic (regulation of cytokines and mediators, nucleus-signalling) and proteomic (structural and functional modification of proteins) high throughput methods.

Actually these methods aiming for understanding the underlying mechanisms will predominantly make use of modern nanotechnology, such as gene and protein interactions (and protein array chip technology specifically designed to screen nanoparticle-gene). This genetic and proteomic information will provide guidance to the next step of assessment. As a result a second series of biologically and toxicologically more relevant tests aimed toward specific reactions may become necessary—first on cells, followed by multi-cell models and *in vivo* animal models which finally may require clinical phase trials in case of medicinal nanoparticles.

From this structured approach, life-cycle-specific recommendations for regulation will be derived for each nanoparticle. Such an approach would even allow the use of a nanoparticle which may have an elevated risk at some stage of its life, but an overall risk–benefit analysis may still be in support of its use including specific prevention measures when they become necessary, e.g. application of nanoparticles in medicine. In fact, controlled application of new emerging nanoparticles will require knowledge-based development, production and application including sustainable risk assessment prior to their widespread use. The public perception is an important factor for the future development of nanosciences and nanotechnologies. Efforts need to be undertaken to convince the public of the beneficial potential of nanosciences and nanotechnologies. It would be a tragedy if a major accident would jeopardize in part or even at large the development of this future technology with its splendid prognoses.

PLANT AND MICROBES AS NANOFACTORIES

There has been an extraordinary growth in nanoscience and technology in recent years, mainly due to both the development of new techniques to synthesize nanomaterials and the accessibility of tools for the classification and manipulation of nanoparticles. The most advanced techniques for the synthesis of nanostructured surfaces include atomic manipulation with scanning probe methods, self-organized growth, and the controlled deposition of nanoclusters from the gas phase. Recently, the utilization of biological systems has emerged as a novel method for the synthesis of nanoparticles. For example, dead biomasses of oat and wheat have shown the ability to create different types of gold nanoparticles when exposed to Au(III)-rich aqueous solutions. Furthermore, alfalfa biomass has proven to bind Au(III) and form nanoparticles when exposed to gold solutions.

There have been impressive developments in the field of nanotechnology in the recent past, with numerous methodologies formulated to synthesize nanoparticles of particular shape and size depending on specific requirements. Currently, there is a growing need to develop environmentally benign nanoparticle synthesis processes that do not use toxic chemicals in the synthesis protocol. As a result, researchers in the field of nanoparticle synthesis and assembly have turned to biological systems for inspiration.

Recently biosynthetic methods employing either biological microorganisms or plant extracts have emerged as a simple and viable alternative to chemical synthetic procedures and physical methods. Following the initial report on intracellular silver nanoparticle formation in *Pseudomonas stutzeri*, many reports on syntheses of metal and semiconductor nanoparticles using fungi or bacteria have appeared. Instead of following some eco-toxic chemical and physical methods to synthesize nanoparticles, eco-friendly "green" methods for the synthesis of noble metal nanoparticles have identified fungi, actinomycetes, and plant extracts for the synthesis of silver and gold nanoparticles. Recently, excellent shape-selective formation of single crystalline triangular gold nanoparticles was observed using the extract of the lemongrass plant (*Cymbopogon flexuosus*). These nanostructures possess a strong near-infrared (NIR) absorbance that could be easily tuned by modifying the experimental conditions. The NIR absorbing properties of the biogenic gold nanotriangles were used to design simple optical coatings for architectural applications.

There are recent reports on the biological synthesis of single crystalline triangular gold nanoparticles in high yield by the reaction of aqueous chloroaurate ions with the extract of the *Aloe vera* plant. This plant has been used in many medical applications as a result of its antipyretic, antioxidative, and cathartic properties. By varying the percentage of the extract in the reaction medium, the percentage of gold nanotriangles to spherical particles as well as the size of the nanotriangles can be modulated, leading to significant control over the optical properties of the nanoparticulate solution. Slow reduction of the aqueous gold ions along with the shape-directing effects of the constituents of the *Aloe vera* extract play a key role in the formation of the gold nanotriangles. On the other hand, reaction of *Aloe vera* extract with aqueous silver ions yields only spherical nanoparticles.

This is not surprising given that many organisms, both unicellular and multicellular, are known to produce inorganic materials either intracellularly or extracellularly. Some well known examples of microorganisms synthesizing inorganic materials include magnetotactic bacteria (which synthesize magnetite nanoparticles), diatoms (which synthesize siliceous materials), and S-layer bacteria (which produce gypsum and calcium carbonate layers). The secrets gleaned from nature have led to the development of biomimetic approaches to the growth of advanced nanomaterials. Even though many biotechnological applications such as remediation of toxic metals employ microorganisms such as bacteria and yeast (the detoxification often occurring via reduction of the metal ions/formation of metal sulphides), it is only relatively recently that materials scientists have been viewing with interest such microorganisms as possible eco-friendly nanofactories. Different species of bacteria have been shown to intracellularly synthesize metal nanoparticles and their alloys.

Prokaryotic organisms (bacteria), eukaryotic organisms such as fungi may also be used to grow nanoparticles of different chemical compositions and sizes including quantum dots of the technologically important CdS by enzymatic processes. Already, there are reported stories of success in the synthesis of fairly monodisperse gold nanoparticles of 8 nm average size using the alkalothermophilic (extremophilic) actinomycete *Thermomonospora* sp., which is currently difficult to achieve through biological means. As can be seen from the above, the use of microorganisms in the deliberate and controlled synthesis of nanoparticles is a relatively new and exciting area of research with considerable potential for development.

While microorganisms such as bacteria, actinomycetes, and fungi continue to be investigated in metal nanoparticle synthesis, the use of parts of whole plants in similar nanoparticle synthesis methodologies is an exciting possibility that is relatively unexplored and underexploited. Even though the gold nanoparticles are considered biocompatible, chemical synthesis methods may still lead to the presence of some toxic chemical species adsorbed on the surface that may have adverse effects in medical applications. Synthesis of nanoparticles using microorganisms or plants can potentially eliminate this problem by making the nanoparticles more biocompatible. Using plants for synthesis of nanoparticles could be advantageous over other environmentally benign biological processes by eliminating the elaborate process of maintaining cell cultures. It can also be suitably scaled up for large-scale synthesis of nanoparticles. Recently, Jose-Yacaman and co-workers demonstrated the synthesis of gold and silver nanoparticles within live alfalfa plants by gold and silver ion uptake, respectively, from solid media. In a related report, agricultural biomass was used to reduce $Cr_{(VI)}$ to $Cr_{(III)}$ ions, indicating that biological methods can be very efficient in decontaminating polluted waters and soil polluted with heavy metal ions.

Some work has been reported work on the synthesis of pure metallic nanoparticles of silver and gold by the reduction of aqueous Ag^+ and $AuCl_4^-$ ions and also the synthesis of bimetallic core-shell nanoparticles of gold and silver by simultaneous reduction of aqueous Ag^+ and $AuCl_4^-$ ions with the broth of neem leaves (*Azadirachta indica*). Through an elaborate screening process involving a number of plants, neem leaves were found to be potential candidates for rapid synthesis of silver and gold nanoparticles.

At high concentration of neem leaf broth, a process for the rapid synthesis of stable silver, gold, and bimetallic Au/Ag core-shell nanoparticles was observed. The flavanone and terpenoid constituents of the leaf broth are believed to be the surface active molecules stabilizing the nanoparticles. The formation of pure metallic and bimetallic nanoparticles by reduction of the metal ions is possibly facilitated by reducing sugars and/or terpenoids present in the neem leaf broth. The present study opens up a new possibility of very conveniently synthesizing bimetallic core-shell nanoparticles using natural products. Achievement of such rapid time scales for synthesis of metallic nanoparticles contributes to an increase in the efficiency of synthetic procedures using environmentally benign natural resources as an alternative to chemical synthesis protocols. Issues such as monodispersity and shape selectivity for obtaining phase pure monodisperse nanoparticles are yet to be addressed and focused on.

Bacteria in Nanoparticle Synthesis

Among the microorganisms, prokaryotic bacteria have received the most attention in the area of biosynthesis of nanoparticles. Early studies reveal that *Bacillus subtilis* 168 is able to reduce Au^{3+} ions to produce octahedral gold particles of nanoscale dimensions (5–25 nm) within bacterial cells by incubation of the cells with gold chloride under ambient temperature and pressure conditions. Organic phosphate compounds play a role in the *in vitro* development of octahedral Au, possibly as bacteria–Au complexing agents. Fe(III)-reducing bacteria *Shewanella algae* can reduce Au(III) ions in anaerobic environments. In the presence of *S. algae* and hydrogen gas, the Au ions are completely reduced, which results in the formation of 10–20 nm gold nanoparticles. It is already established that silver is highly toxic to most microbial cells. Nonetheless, several bacterial strains are reported as silver-resistant and may even accumulate silver at the cell wall to as much as 25% of the dry weight biomass, thus suggesting their use for the industrial recovery of silver from ore material. The silver-resistant bacterial strain *Pseudomonas stutzeri* AG259 accumulates silver nanoparticles, along with some silver sulphide, in the cell where particle size ranges from 35 to 46 nm. Larger particles are formed when *P. stutzeri* AG259, isolated from a silver mine, is placed in a concentrated aqueous solution of silver nitrate (50 mM). Nanoparticles of well-defined size, ranging from a few to 200 nm or more, and distinct morphology are deposited within the periplasmic space of the bacteria. Cell growth and metal incubation conditions may be the reasons for the formation of different particle sizes. The exact reaction mechanisms leading to the formation of silver nanoparticles by this species of silver-resistant bacteria is yet to be elucidated. The ability of microorganisms to grow in the presence of high metal concentrations might result from specific mechanisms of resistance. Such mechanisms include the following: efflux systems, alteration of solubility and toxicity by changes in the redox state of the metal ions, extracellular complexation or precipitation of metals, and the lack of specific metal transport systems. Biocomposites of nanocrystalline silver and the bacteria may be thermally treated to yield a carboneous (cermet) nanomaterial with interesting optical properties for potential application in functional thin-film coating. These properties can be modified by varying the silver loading factor. This type of carboneous material is composed primarily of graphite carbon and up to 5% by weight (of the dry biomass) of silver.

Bacteria not normally exposed to large concentrations of metal ions may also be used to grow nanoparticles. The exposure of *Lactobacillus* strains, which are present in buttermilk, to silver and gold ions resulted to the large-scale production of metal nanoparticles within the bacterial cells. Moreover, the exposure of lactic acid bacteria present in the whey of buttermilk to mixtures of gold and silver ions can be used to grow alloy nanoparticles of gold and silver. In addition to gold and silver nanoparticles, there is much attention in the development of protocols for the synthesis of semiconductors (the so-called quantum dots) such as CdS, ZnS, and PbS. These luminescent quantum dots are emerging as a new class of materials for biological detection and cell imaging, based on the conjugation of semiconducting quantum dots and biorecognition molecules. *Clostridium thermoaceticum* precipitates CdS at the cell surface as well as in the medium from $CdCl_2$ in the presence of cysteine hydrochloride in the growth medium. Most probably, cysteine acts as the source of sulphide. When *Klebsiella aerogenes* is exposed to Cd^{2+} ions in the growth medium, 20–200 nm CdS is formed on the cell surface. The formation of CdS is confirmed by quantitative energy-dispersive X-ray analysis. The buffer composition of the growth medium plays an important role in forming the cadmium sulphide crystallites. Intracellular CdS nanocrystals, composed of a wurtzite crystal phase, are formed when *E. coli* is incubated with cadmium chloride and sodium sulphide. Nanocrystal formation varies dramatically depending on the growth phase of the cells and increases about 20-fold in *E. coli* grown in the stationary phase as compared with that grown in the late logarithmic phase. In a remarkable investigation, Labrenz *et al.* have shown that spherical aggregates of 2–5 nm diameter sphalerite (ZnS) particles are formed within natural biofilms dominated by sulphate-reducing bacteria of the family Desulfobacteriaceae. A combination of geochemical and microbial processes leads to ZnS biomineralization in a complex natural system. It is appropriate to mention that the concentration of Zn can be significantly reduced to below-acceptable levels for drinking water with the use of this method. Magnetic Fe sulphide nanoparticles are synthesized by using sulphate-reducing bacteria where particles having a size of a few nanometres are formed on the surface, and the magnetic mineral is separated from the solution by a high-gradient magnetic field of 1 T. Bacterially produced iron sulphide is an adsorbent for a wide range of heavy metals and some anions. High adsorption of radio-486-active ions occurs due to high surface area (400–500 m^2/g).

It may provide a suitable matrix for the long-term safe storage of a number of ions vital to the nuclear industry, particularly the pertechnetate ion (TcO_4^-). Magnetite is a common product of bacterial iron reduction and could be a potential physical indicator of biological activity in geological settings.

Single-domain tiny magnetic particles (<12 nm), which exhibit octahedral shapes, are formed exclusively outside of the bacterial cells by a thermophilic fermentive bacterial strain TOR-39. Transition metals such as Co, Cr, and Ni may be substituted for magnetic crystals biosynthesized in the thermophilic iron-reducing bacteria *Thermoanaerobacter ethanolicus* (TOR-39) by way of the electrochemical process. The mineralization processes are highly controlled by the magnetotactic bacteria, leading to the formation of uniform, species-specific magnetic nanoparticles. Sometimes, the particles are assembled into single or multiple chains and anchored inside the cell, enabling the bacteria to passively orient themselves along geomagnetic field lines. Interestingly, the magnetotactic bacteria *Magnetospirillum magnetotacticum* produce single-domain magnetic crystals (Fe_3O_4) that are subsequently assembled into folded chain and flux-closure ring morphologies. The magnetic crystals with large magnetic moments, when constrained to lie on a two-dimensional surface, are responsible for the head-to-tail assembly. Based on the magnetization measurements and the magnitude of the magnetization field, it is established that biogenic magnetite nanoparticles are not superparamagnetic. Magnetic nanoparticles are also assembled into ordered structures when the motion of the magnetic bacteria M. *magnetotacticum* (MS-1) is controlled by applying a magnetic field. After assembling the bacteria with microelectromagnets, the cellular membranes of the bacteria are removed by cell lysis to leave the biogenic magnetic nanoparticles at desired locations.

Different patterns of magnetic structures are observed after removing the cellular membrane of trapped MS-1 bacteria. These types of ordered magnetic structures could serve as a system for studying the interactions between closely spaced magnetic nanoparticles. Thus, a different approach, the combination of biomineralization and micromanipulation, can be a new procedure for growing and assembling nanoparticles into customized structures. All magnetotactic bacteria contain magnetosomes, which are intracellular structures comprising magnetic iron mineral crystals enveloped by a membrane vesicle. The magnetosome membrane (MM) is most likely a structural entity that anchors the crystal at particular locations in the cell, as well as the locus of biological control over the nucleation and growth of the magnetosome crystals. Knowledge of biochemical and genetic controls on magnetic production is essential to understanding how the magnetotactic bacteria produce magnetosomes and organize them in chains. The biomineralization of magnetosome particles is achieved by a complex mechanism that involves the uptake and accumulation of iron and the deposition of the mineral particle with a specific size and morphology within a specific section provided by the MM.

Since the MM is thought to be of paramount importance in magnetosome formation, researchers have focused on the role of MM proteins, which occur in the MM but not in the soluble (periplasmic or cytoplasmic) fraction or in the cytoplasmic or outer membranes, in magnetosome synthesis. The mam (MM) genes appear to be conserved in a large gene cluster within several magnetotactic bacteria (*Magnetospirillum* species and strain MC-1) and may be involved in magnetic biomineralization. The monodispersity of the silver/gold nanoparticles produced either intra- or extracellularly by the above mentioned methodology is not very high and far inferior to that obtained through conventional chemical methods. More by chance than by design, it is observed that alkalothermophilic (extremophilic) actinomycete, *Thermomonospora* sp., when exposed to gold ions, reduces the metal ions extracellularly, yielding gold nanoparticles with much polydispersity. A complete reduction of the 10^{-3} M aqueous $HAuCl_4$ solution at pH 9.0 and 50°C results to spherical and reasonably monodisperse nanoparticles (8 nm). In contrast, intracellular synthesis of gold nanoparticles occurs in alkalotolerant actinomycete *Rhodococcus* sp., where particles are more concentrated on the cytoplasmic membrane than on the cell wall. There are very few reports regarding alga-mediated synthesis of nanoparticles to date. Suspension of dried cells of alga *Chlorella vulgaris* in the $HAuCl_4$ solution accumulates elemental gold in the cells. Phytochelation (PC)-coated CdS nanocrystallites formed in a marine phytoplanktonic alga *Phaeodactylum tricornutum* in response to Cd.

Yeast in Nanoparticle Synthesis

It has long been recognized that among the eukaryotes, yeasts are explored mostly in the biosynthesis of the semiconductor nanoparticles. Exposure of *Candida glabrata* to Cd^{2+} ions leads to the intracellular formation of CdS quantum dots. The synthesis of PC is activated in the presence of Cd. The structure of PC involves a repeating sequence of γ-glutamyl–cysteine pairs to give polypeptides the general formula $(\gamma\text{-Glu-Cys})_n$ Gly, with *n* values commonly ranging from 2 to 6. They bind Cd ions immediately, forming Cd–PC complexes that are transported into the vacuole. Then, the complex is degraded and the nanoparticles are formed. *Torulopsis* sp., which was found in an extensive screening program, is capable of synthesizing PbS nanocrystals intracellularly when challenged with Pb^{2+}.

Crystallites, which are extracted from the biomass by freeze thawing, exhibit a sharp absorption maximum at ~330 nm and are 2–5 nm in size. Intensive research activity is currently in progress to fabricate electronic devices using organic materials due to their flexibility, easy processing, and large quantum efficiency for light emission. To improve their stability, efficiency, and color tunability for diverse applications, composites of organic materials with nanoparticles, porous silicon, etc., have been explored. So far, chemically synthesized nanoparticles have been used for device applications, but biogenic nanoparticles are yet to be explored for the same applications. CdS quantum dots synthesized intracellularly in *Schizosaccharomyces pombe* yeast cells exhibit ideal diode characteristics. Biogenic CdS nanoparticles in the size range 1–1.5 nm have been used in the fabrication of a heterojunction with poly(p-phenylenevinylene). Such a diode exhibits an approximately 75-mA/cm^2 current in the forward bias mode at 10 V, while breakdown occurred at ~15 V in the reverse direction. Though yeast has been used to synthesize intracellular nanoparticles for several years, very recently, silver nanoparticles have been synthesized extracellularly by a silver-tolerant yeast strain, MKY3. Particles with a 2–5 nm size range are formed when challenged with Ag^+ ions in the log phase of growth.

Fungi in Nanoparticle Synthesis

The use of fungi in the synthesis of nanoparticles is a relatively recent addition to the list of microorganisms. The use of fungi is potentially exciting since they secrete large amounts of enzymes and are simpler to deal with in the laboratory. However, the genetic manipulation of eukaryotic organisms as a means of overexpressing specific enzymes identified in nanomaterial synthesis would be much more difficult than that in prokaryotes. An extensive screening process resulted to two genera, which, when challenged with aqueous metal ions such as $AuCl_4^-$ and Ag^+, yielded large quantities of metal nanoparticles either extracellularly or intracellularly. The appearance of a distinctive purple color in the biomass of *Verticillium* sp. after exposure to the 10^{-4} M $HAuCl_4$ solution indicates the formation of gold nanoparticles intracellularly and can clearly be seen in the UV–Visible absorption spectrum recorded from the gold-loaded biomass as a resonance at ~550 nm. In the low magnification TEM image, a number of *Verticillium* sp. cells can be seen; small particles are organized on the walls of the cells and larger particles are observed within the cells. At higher magnification, the 5–200 nm sized nanoparticles with an average size of 20 ± 8 nm are clearly seen populating both the cell wall and the cytoplasmic membrane of the single cell of the fungi.

Furthermore, the powder diffraction pattern recorded from the biofilm indicates the crystalline nature of gold nanoparticles. The exposure of *Verticillium* sp. to silver ions resulted to a similar intracellular growth of silver nanoparticles. The exact mechanism leading to the intracellular formation of gold and silver nanoparticles by *Verticillium* sp. is not fully understood at the moment. Since the nanoparticles are formed on the surface of the mycelia and not in the solution, it is thought that the first step involves the trapping of the metal ions on the surface of the fungal cells possibly via electrostatic interaction between the ions and the negatively charged carboxylate groups in the enzymes present in the cell wall of the mycelia. Thereafter the ions are reduced by enzymes present in the cell wall leading to the formation of the nuclei, which subsequently grow through the further reduction of metal ions and accumulation of these nuclei. The ability of *Verticillium* sp. cells to multiply after exposing to metal ions proves the capability of using microorganisms in the synthesis of nanomaterials. From the application point of view, it would be imperative to harvest the metal nanoparticles formed within the fungal biomass. It is possible to release the intracellular silver and gold nanoparticles via ultrasound treatment of the biomass–nanoparticles composite or via reaction with suitable detergents. Nonetheless, it would be far more practical if the metal ions exposed to the fungus could be reduced outside the fungal biomass, leading to the formation of metal nanoparticle in the solution. Quite surprisingly, the plant pathogenic fungal strain *Fusarium oxysporum* behaved considerably differently; the reduction of the metal ions occurred extracellularly, resulting in the rapid formation of highly stable gold and silver nanoparticles of 2–50 nm dimensions. Moreover, the aqueous extract of the fungal biomass can reduce gold and silver ions to the corresponding nanoparticles. Most probably, the reduction of the $AuCl_4$ and Ag^+ ions occurs due to reductases released by the fungus into the solution, thus opening up a novel fungal/enzyme-based *in vitro* approach to nanomaterials.

The long-term stability of the nanoparticles in the solution may be due to the stabilization of proteins, like cysteine. Apart from individual metal nanoparticles, the bimetallic Au–Ag alloy can be synthesized by *F. oxysporum*. Very recently, it has been shown that when the biomass of *F. oxysporum* is exposed to equimolar solutions of $HAuCl_4$ and $AgNO_3$, highly stable Au–Ag alloy nanoparticles of varying mole fractions can be achieved. Variation in the amount

TABLE 6.1 ORGANISMS IN NANOPARTICLE SYNTHESIS

Organisms	Nanoparticles
Bacteria	
Bacillus subtilis	Gold
Shewanella algae	Gold
Pseudomonas stutzeri	Silver
Lactobacillus sp.	Gold, silver, Au-Ag alloy
Clostridium thermoaceticum	Cadmium sulphide
Klebsiella aerogenes	Cadmium sulphide
Escherichia coli	Cadmium sulphide
Desulfobacteriaceae spp.	Zinc sulphide
Thermoanaerobacter ethanolicus	Magnetite
Magnetospirillium magnetotacticum	Magnetite
Thermomonospora sp.	Gold
Rhodococcus sp.	Gold
Chlorella vulgaris	Gold
Phaeodactylum tricornutum	Cadmium sulphide
Yeast	
Candida glabrata	Cadmium sulphide
Torulopsis sp.	Lead sulphide
Schizosaccharomyces pombe	Cadmium sulphide
MKY3	Silver
Fungi	
Verticillium sp.	Gold, silver
Fusarium oxysporum	Gold, silver, Au-Ag alloy, cadmium sulphide, zirconia
Colletotrichum sp.	Gold

of biomass reveals that secreted cofactor NADH plays an important role in determining the composition of Au-Ag alloy nanoparticles. An almost uniform contrast for each particle in the TEM analysis suggests that the electron density is homogeneous within the volume of the particle, which is unlikely in the case of the core/shell structure. Even more exciting is the finding that the exposure of *F. oxysporum* to the aqueous $CdSO_4$ solution yields CdS quantum dots extracellularly. The particles are reasonably monodisperse and range in size from 5 to 20 nm. X-ray diffraction analysis of a film of the particles formed on a Si wafer clearly shows that the particles are nanocrystalline with Bragg reflection, which is characteristic of hexagonal CdS. Reaction of the fungal biomass with the aqueous $CdNO_3$ solution for an extended period of time does not yield CdS nanoparticles, indicating the possibility of the release of a sulfate reductase enzyme into the solution. Polyacrylamide gel electrophoresis indicates the presence of at least four protein bands in the aqueous extract of the fungal biomass. Reaction of the protein extract after dialysis (using a dialysis bag with a 3-kDa molecular weight cutoff) with the $CdSO_4$ solution does not yield CdS nanoparticles. However, the addition of ATP and NADH to the dialysate restores the CdS formation capability of the protein extract. It is believed that the same proteins are also responsible for the reduction of gold and silver ions. Another significant application of this fungus is in the synthesis of zirconia nanoparticles, which are a technologically promising oxide.

The challenging of the aqueous solution of K_2ZrF_6 to the fungus *F. oxysporum* results to protein-mediated extracellular hydrolysis of the zirconium hexafluoride anions and room temperature formation of crystalline zirconia nanoparticles. An endophytic fungus (*Colletotrichum* sp.) growing in the geranium leaves, when exposed to aqueous chloroaurate ions, produces gold nanoparticles of rodlike and prismatic morphology.

CONCLUDING REMARKS

While the science of nanomaterials and human health impact is maturing, it is still at a stage of raising many more questions than answers. Current research demonstrates that some engineered nanomaterials can behave differently in the body than more conventional materials, and may present a health risk that is not captured within established risk assessment paradigms. There is little published research on the importance of characteristics unique to the nanoscale (as opposed to properties that scale with size), and most risk-based research appears to be focused on first-generation engineered nanomaterials, despite the concurrent development of second- and third-generation technologies. Against this background, significant investments are being made in nanotechnology, commercial products are anticipated to increase, public awareness of the potential benefits and risks is growing, and discussions are beginning on oversight of the technology.

Quantitative risk assessment remains difficult for engineered nanomaterials. It is reasonable to speculate that there will be risks, and that conventional risk assessment paradigms will not always suffice. However, specific information on hazard, exposure, dose, response, and other compartments within risk assessment frameworks is lacking. At the same time, an increasingly influential public is shaking the reliance on science-based risk governance alone. Failures of public trust in technologies such as nuclear power and genetically modified foods have demonstrated the power of perceived risk in determining success or failure.

If oversight of nanotechnology is to nurture beneficial technologies rather than stifle them, it will be necessary to develop appropriate ways of working within a framework of scientifically sound information and public perception. Recent research into public perceptions has indicated enthusiasm over the potential uses of nanotechnology, but concern over the ability of industry and government to regulate it. A recent independent report acknowledges that regulation of nanotechnology will be difficult, but concludes that if nothing specific is done to manage the potential adverse effects of nanotechnology, "the public potentially would be left unprotected, the government would struggle to apply existing laws to a technology for which they were not designed, and industry would be exposed to the possibility of public backlash, loss of markets, and potential financial liabilities".

Despite current risk-based uncertainty and the complexity of overseeing the emergence of "safe" nanotechnologies, one certainty is the need for further research and better data. A recent analysis of current environment, safety, and health-impact research indicated that there is a significant amount of preemptive work addressing potential risks of engineered nanomaterials. However the same analysis concluded that there is a lack of overall strategy in the current research portfolio, and that without a clear strategic framework, greater resources, and increased partnerships and collaboration, critical questions are unlikely to be answered in a timely manner.

Research community from nanotechnology background strongly believing that "Bionanofactories" will overcome the problem which is coming out with environmental concern. Instead of chemical and physical methods for nanomaterial production, using microbes and plants will help a lot to synthesize the materials in the nano range and in addition, the toxicity of the by-product would be lesser than the others. Some researchers are strongly supporting that there willnot be any release of toxic substances during the nanoparticle synthesis with the help of plant or microbes. The reason is chemicals which will be used in nanoparticles synthesis will get degraded by the enzymatic substances which are produced by the microbes during the time of growth. Plants also by trap the chemical substances within their bodies and use the same as nutritive materials for metabolic processes of their own. This is why microbes are considered as eco-friendly nanoproducers and plants as nanofactories.

CONCEPT CHECK

1. What are the properties of SWNT and MWNT?
2. Give the statement of Paracelsus.
3. Is nanotoxicology based on surface of the nanoparticle or accumulation?
4. Mention the most predominantly used microbes for nanoparticle synthesis.
5. Plant vs Microbes–Which is more convenient for NP synthesis?

MIND TWISTER

1. Shape of the NP is based on the inner anatomic structure of the plant–True or False? Give reasons.
2. Which is more efficient in medical application–plant/microbes producing NP or chemical/physical-based synthesis of NP? Why?

REFERENCES

Baron, P.A. *et al.* Evaluation of aerosol release during the handling of unrefined carbon nanotube material, NIOSH, Cincinnati, OH, (2003). DART-02-191/NTIS PB2003-102401.

Baron, P.A. *et al.* (2001). Nonspherical particle measurements: shape factors, fractals and fibres. In: *Aerosol Measurement. Principles, Techniques and Applications*, 2nd edn. Baron, P.A. and Willeke, K. (eds.). John Wiley and Sons, NY. p.705.

Donaldson, K. *et al.* (2002). Deposition and clearance effects in the lung. *J. Aerosol Med.* 15(2). 213–220.

Lovley, D.R., Stolz, J.F., Nord, G.L. and Philips, E.J.P. (1987). Anaerobic production of magnetite by a dissimilatory iron-reducing microorganism. *Nature.* 330: 252–254.

Maynard, A. D. (2005). Inventory of Research on the Environmental, Health and Safety Implications of Nanotechnology, Woodrow Wilson International Center for Scholars, Project on Emerging Nanotechnologies, Washington DC, USA.

Mukherjee, P., Ahmad, A., Mandal, D., Senapati, S., Sainkar, S.R., Khan, M.I., Ramani, R., Parischa, R., Ajayakumar, P.V., Alam, M., Sastry, M. and Kumar, R. (2001a). Bioreduction of $AuCl^4$ ions by the fungus, *Verticillium* sp. and surface trapping of the gold nanoparticles formed. *Angew Chem. Int. Ed.* 40: 3585–3588.

Senapati, S., Ahmad, A., Khan, M.I., Sastry, M. and Kumar, R. (2005). Extracellular biosynthesis of bimetallic Au–Ag alloy nanoparticles. *Small.* 1: 517–520.

Shankar, S.S., Ahmad, A., Parischa, R. and Sastry, M. (2003). Bioreduction of chloroaurate ions by geranium leaves and its endophytic fungus yields gold nanoparticles of different shapes. *J. Mater. Chem.*13: 1822–1826.

Spring, H. and Schleifer, K.H. (1995). Diversity of magnetotactic bacteria. *Syst. Appl. Microbiol.* 18: 147–153.

Sweeney, R.Y., Mao, C., Gao, X., Burt, J.L., Belcher, A.M., Georgiou, G. and Iverson, B.L. (2004). Bacterial biosynthesis of cadmium sulphide nanocrystals. *Chem. Biol.* 11: 1553–1559.

Tran, C.L. *et al.* (1999). Investigation and prediction of pulmonary responses to dust. Part II. In: Investigations into the pulmonary effects of low toxicity dusts. Parts I and II., Health and Safety Executive UK, Suffolk, UK. Contract Research Report 216/1999.

GLOSSARY

AFM Atomic force microscope (AFM) or scanning force microscope (SFM). It is a very high-resolution type of scanning probe microscopy, invented by Binnig, Quate and Gerber in 1986 with demonstrated resolution of fractions of a nanometre, more than 1000 times better than the optical diffraction limit. Piezoelectric elements that facilitate tiny but accurate and precise movements on (electronic) command enable the very precise scanning.

Angstrom A unit of measurement which is equivalent to one ten-billion of a metre (0.0000000001 m) abbreviated as A, and is used mainly to represent the wavelengths of visible light There are 10 million angstroms in a centimetre.

ANN Artificial neural network. It is usually called "neural network" (NN), and is a mathematical or computational model that tries to simulate the structure and/or functional aspects of biological neural networks.

Atom A basic unit of matter consisting of a dense, central nucleus surrounded by a cloud of negatively charged electrons. The atomic nucleus contains a mix of positively charged protons and electrically neutral neutrons (except in the case of hydrogen-1, which is the only stable nuclide with no neutron). The electrons of an atom are bound to the nucleus by the electromagnetic force. Likewise, a group of atoms can remain bound to each other, forming a molecule.

Biopolymers A major but defining difference between polymers and biopolymers can be found in their structures. Biopolymers often have a well defined structure, though this is not a defining characteristic (e.g. lignocellulose). The exact chemical composition and the sequence in which these units are arranged is called the primary structure, in the case of proteins. Many biopolymers spontaneously fold into characteristic compact shapes, which determine their biological functions and depend in a complicated way on their primary structures.

Bottom-up approach It is based on piecing together systems to give rise to grander systems, thus making the original systems sub-systems of the emergent system. In a bottom-up approach the individual base elements of the system are first specified in great detail. These elements are then linked together to form larger subsystems, which then in turn are linked, sometimes in many levels, until a complete top-level system is formed.

Catalyst Substance (composite or element) that can speed up a chemical reaction, while remaining unaltered itself. This process is called catalysis. Many catalysts act by increasing the surface area, thus allowing two or more chemical reagents to come together and unite or separate. Catalysts do not alter the final energy balance of the chemical reaction; they simply allow balance to be achieved at a greater or lesser speed.

Chemical Vapour Deposition (CVD) A technique used to deposit coatings, where chemicals are first vaporized, and then applied using an inert carrier gas such as nitrogen.

Chemotherapy Most commonly, chemotherapy acts by killing cells that divide rapidly, one of the main properties of cancer cells. This means that it also harms cells that divide rapidly under normal circumstances: cells in the bone marrow, digestive tract and hair follicles; these results in the most common side effects of chemotherapy. In a general sense, chemotherapy refers to treatment of disease by chemicals that kill cells, both good and bad, but specifically those of cancerous cells. The first modern chemotherapeutic agent was Paul Ehrlich's arsphenamine, an arsenic compound discovered in 1909 and used to treat syphilis.

CNTs Carbon nanotubes.They are allotropes of carbon with a cylindrical nanostructure. These cylindrical carbon molecules have novel properties that make them potentially useful in many applications in nanotechnology, electronics, optics and other fields of materials science, as well in the field of architecture. They exhibit extraordinary strength and unique electrical properties, and are efficient thermal conductors. Nanotubes are members of the fullerene structural family, which also includes the spherical buckyballs.

Colloidal particle A small amount of matter having size typical for colloids and with a clear phase boundary (phase colloids). A group of such particles are aggregate, or agglomerate and being a macromolecule or a molecular aggregate (e.g. micelle) said to be colloidal particle. A **colloidal crystal** is an ordered array of particles, analogous to a standard crystal whose repeating subunits are in the size of atoms or molecules.

Composite materials Engineered materials made from two or more constituent materials with significantly different physical or chemical properties which remain separate and distinct on a macroscopic level within the finished structure.

Crystallinity The degree of structural order in a solid. In a crystal, the atoms or molecules are arranged in a regular, periodic manner and the degree of crystallinity has a big influence on hardness, density, transparency and diffusion.

Dendrimer The word dendrimer comes from the Greek *dendron* ("tree") and the suffix *mer* ("segment"). Dendrimers are built in the nanoscale from layers of monomers (a single molecule with the capacity to combine with identical or similar molecules) to create a tree structure. The first layers of monomers, the nucleus, is known as "generation 0".

DPN Dip pen nanolithography. It is a scanning probe lithography technique where an atomic force microscope tip is used to transfer molecules to a surface via a solvent meniscus. This technique allows surface patterning on scales of under 100 nanometres. DPN is the nanotechnology analog of the dip pen (also called the quill pen), where the tip of an atomic force microscope cantilever acts as a "pen," which is coated with a chemical compound or mixture acting as an "ink," and put in contact with a substrate. DPN enables direct deposition of nanoscale materials onto a substrate in a flexible manner. Applications of this technology currently range through chemistry, materials science, and the life sciences, and include such work as ultrahigh-density biological nanoarrays, additive photomask repair, and brand protection for pharmaceuticals.

DSC Differential scanning calorimetry. It is a thermoanalytical technique in which the difference in the amount of heat required to increase the temperature of a sample and reference are measured as a function of temperature. Both the sample and reference are maintained at nearly the same temperature throughout the experiment.

Electrospinning The process that uses an electrical charge to draw very fine (typically on the micro- or nanoscale) fibres from a liquid. It is a non-invasive process and does not require the use of coagulation chemistry or high temperatures to produce solid threads from solution. This makes the process particularly suited to the production of fibres using large and complex molecules.

Focused Ion Beam (FIB) The system that uses a beam of gallium ions to launch localized ionic attacks on materials and localized deposits of various materials. It makes it possible to view and create three-dimensional structures, controlling the processes with a precision in the tens of nanometres. An FIB system operates in much the same way as a scanning electron microscope (SEM).

Fullerene Any molecule composed entirely of carbon, in the form of a hollow sphere or tube. Spherical fullerenes are also called buckyballs, and cylindrical ones are called carbon nanotubes or buckytubes. The first fullerene to be discovered, and the family's namesake, was buckminsterfullerene C_{60}, made in 1985 by Robert Curl, Harold Kroto and Richard Smalley. Buckyballs and buckytubes have been the subject of intense research, both for their unique chemistry and for their technological applications, especially in materials science, electronics, and nanotechnology.

Glass transition The transformation of a glass-forming liquid into a glass, which usually occurs upon rapid cooling. It is a dynamic phenomenon occurring between two distinct states of matter (liquid and glass), each with different physical properties. The glass transition temperature T_g is lower than the melting temperature, T_m, due to supercooling. It depends on the time scale of observation which must be defined by convention.

Grain boundary The interface between two grains in a polycrystalline material. Grain boundaries disrupt the motion of dislocations through a material, so reducing crystallite size is a common way to improve strength, as described by the Hall-Petch relationship. Since grain boundaries are defects in the crystal structure, they tend to decrease the electrical and thermal conductivity of the material.

Implant A medical device made to replace and act as a missing biological structure (as compared with a transplant, which indicates transplanted biomedical tissue). The surface of implants that contact the body might be made of a biomedical material such as titanium, silicone or apatite depending on what is the most functional. Some implants are bioactive, such as subcutaneous drug delivery devices in the form of implantable pills or drug-eluting stents. In orthopaedic surgery, implants may refer to devices that are placed over or within bones to hold a fracture reduction.

Integrins Receptors that mediate attachment between a cell and the tissues surrounding it, which may be other cells or the extracellular matrix (ECM). They also play a role in cell signalling and thereby define cellular shape, and mobility, and regulate the cell cycle.

Lab-on-a-chip The integration of microapparatuses and nanofluids holds out the prospect of automating chemistry in a minute system. Through this nano-opening, analyses and experiments can be carried out on the patient without the involvement of the lab technician.

Liposomes A tiny bubble (vesicle) that can be filled with drugs, and used to deliver drugs for cancer and other diseases. Membranes are usually made of phospholipids, which are molecules that have a head group and a tail group. The head is attracted to water, and the tail, which is made of a long hydrocarbon chain, is repelled by water.

Lithography A method for printing using a stone (lithographic limestone) or a metal plate with a completely smooth surface. Invented by Bavarian author Alois Senefelder in 1796, it can be used to print text or artwork onto paper or another suitable material. The word "lithography" also refers to photolithography, a microfabrication technique used to make integrated circuits and microelectromechanical systems.

MEMS Microelectromechanical systems. The system is made up of components between 1 to 100 micrometres in size (i.e., 0.001 to 0.1 mm) and MEMS devices generally range in size from 20 micrometres (20 millionths of a metre) to a millimetre. They usually consist of a central unit that processes data, the microprocessor and several components that interact with the outside such as microsensors.

Micelle An aggregate of surfactant molecules dispersed in a liquid colloid. A typical micelle in aqueous solution forms an aggregate with the hydrophilic "head" regions in contact with surrounding solvent, sequestering the hydrophobic single tail regions in the micelle centre. Micelles are approximately spherical in shape. The shape and size of a micelle is a function of the molecular geometry of its surfactant molecules and solution conditions such as surfactant concentration, temperature, pH, and ionic strength. The process of forming micellae is known as micellization and forms part of the phase behaviour of many lipids according to their polymorphism.

Microemulsions Clear, stable, isotropic liquid mixtures of oil, water and surfactant, frequently in combination with a co-surfactant. The two basic types of microemulsions are direct (oil dispersed in water, o/w) and reversed (water dispersed in oil, w/o). In ternary systems such as microemulsions, where two immiscible phases (water and 'oil') are present with a surfactant, the surfactant molecules may form a monolayer at the interface between the oil and water, with the hydrophobic tails of the surfactant molecules dissolved in the oil phase and the hydrophilic head groups in the aqueous phase.

Micro-encapsulation A process in which tiny particles or droplets are surrounded by a coating to give small capsules many useful properties. In a relatively simplistic form, a microcapsule is a small sphere with a uniform wall around it. The material inside the microcapsule is referred to as the core, internal phase, or fill, whereas the wall is sometimes called a shell, coating, or membrane. Most microcapsules have diameters between a few micrometres and a few millimetres.

Microfluidics The science of designing, manufacturing, and formulating devices and processes that deal with volumes of fluid on the order of nanolitres (symbolized nl and representing units of 10^{-9} litre) or picolitres (symbolized "pl" and representing units of 10^{-12} litre).

Microlithography and Nanolithography Generally this refers to lithographic patterning methods capable of structuring material on a fine scale. Electron beam lithography is capable of much higher patterning resolution (sometimes as small as a few nanometres). If the features are smaller than 10 micrometres, they are considered microlithographic, and if the features are smaller than 100 nanometres, they are considered nanolithographic.

Molecular recognition A chemical term referring to processes in which molecules adhere in a highly specific way, forming a larger structure; an enabling technology for nanotechnology.

Molecule An electrically neutral group of at least two atoms in a definite arrangement held together by very strong (covalent) chemical bonds. In organic chemistry and biochemistry, the term molecule is used less strictly and also is applied to charged organic molecules and biomolecules. In the kinetic theory of gases the term molecule is often used for any gaseous particle regardless of its composition. A molecule may consist of atoms of a single chemical element, as with oxygen (O_2), or of different elements, as with water (H_2O).

MWNT Multi-walled nanotubes. Most of these nanotubes consist of multiple rolled layers (concentric tubes) of graphite. There are two models which can be used to describe the structures of multi-walled nanotubes. In the Russian Doll model, sheets of graphite are arranged in concentric cylinders and in the Parchment model, a single sheet of graphite is rolled in around itself.

Nanocomposite A multiphase solid material where one of the phases has one, two or three dimensions of less than 100 nanometres (nm) which include porous media, colloids, gels and copolymers.

Nanocomposites A lighter, harder and more rigid material at low temperatures than traditional thermoplastics. Nanocomposites are created by introducing a solid material into a plastic resin to make it stronger. Because there is less additive material, they recycle better than other thermoplastics.

Nanolithography A branch of nanotechnology, which deals with the study and application of fabrication of nanoscale structures like semiconductor circuits. As of 2007, nanolithography is a very active area of research in academia and in industry. Generally it refers to the fabrication of nanometre-scale structures, meaning patterns with at least one lateral dimension between the size of an individual atom and approximately 100 nm.

Nanomedicine The branch of nanotechnology that will allow diseases to be cured from the inside of the body at a cellular or molecular scale. It is one of the most promising areas of potential new developments in technological applications for medicine.

Nanoparticle Particle with a size less than 100 nm. The exceptional properties of nanoparticles are largely due to the material on the surface, which does not have the same properties as the same materials in larger volumes. Because they are so small they can reflect light instead of absorbing it. In general, biomedicine and biotechnology are two promising fields in terms of their potential applications.

Nanopores Involves squeezing a DNA sequence between two oppositely charged fluid reservoirs, separated by an extremely small channel. Nanoscopic pores found in purpose-built filters, sensors, or diffraction gratings make them function better.

Nanoporous materials Materials manipulated at the nanoscale; the size of their pores has been altered to almost perfect levels. It will also be possible to manipulate their physical and chemical characteristics. They will be used especially in medicine and pharmacy.

Nanoprobe Nanoscale machines used to diagnose, image, report on, and treat disease within the body.

Nanosensors A probe or sensor with nanometric accuracy. Such probes are currently experiencing rapid development in nanotechnology. Because of their many applications, nanosensors can also be said to have the potential to trigger revolutionary changes in nearly all areas of science and technology.

Nanoshells A type of spherical nanoparticles consisting of a dielectric core which is covered by a thin metallic shell (usually gold). They can absorb or scatter light at virtually any wavelength. Nanoshells act as an amazingly versatile optical component on the nanometre scale. They may provide a whole new approach to optical materials and components.

Nanotoxicology A branch of bionanoscience, which deals with the study and application of toxicity of nanomaterials. Because of quantum size effects and large surface area, nanomaterials have unique properties compared with their larger counterparts. Nanotoxicological studies are intended to determine whether and to what extent these may pose a threat to the environment and to human beings. For instance, diesel nanoparticles have been found to damage the cardiovascular system in a mouse model.

Nanowires Semiconductor nanowires are one-dimensional structures with unique electrical and optical properties, which are used as building blocks in nanoscale devices.

NEMS Nanoelectromechanical systems. It is used to describe devices integrating electrical and mechanical functionality on the nanoscale. NEMS form the logical next miniaturization step from so-called microelectromechanical systems, or MEMS devices. NEMS typically integrate transistor-like nanoelectronics with mechanical actuators, pumps, or motors, and may thereby form physical, biological, and chemical sensors.

Neurotransmitters These are endogenous chemicals which relay, amplify, and modulate signals between a neuron and another cell. Neurotransmitters are packaged into synaptic vesicles that cluster beneath the membrane on the presynaptic side of a synapse, and are released into the synaptic cleft, where they bind to receptors in the membrane on the postsynaptic side of the synapse. Release of neurotransmitters usually follows arrival of an action potential at the synapse, but may follow graded electrical potentials.

Optoelectronics Electrical-to-optical or optical-to-electrical transducers, or instruments that use such devices in their operation. It is the study and application of electronic devices that source, detect and control light, usually considered a sub-field of photonics. In this context, light often includes invisible forms of radiation such as gamma rays, X-rays, ultraviolet and infrared, in addition to visible light.

PDT Photodynamic therapy. It is an advanced medical technology started in the 1980s at several institutions throughout the world, and is a third-level treatment for cancer involving three key components: a photosensitizer, light, and tissue oxygen.

Photosensitizer A chemical compound that can be excited by light of a specific wavelength. This excitation uses visible or near-infrared light. In photodynamic therapy, either a photosensitizer or the metabolic precursor of one is administered to the patient. The tissue to be treated is exposed to light suitable for exciting the photosensitizer.

Polymer A large molecule (macromolecule) composed of repeating structural units called monomers which are typically connected by covalent chemical bonds.

Quantum wire Another form of quantum dot, but unlike the single-dimension "dot", a quantum wire is confined only in two dimensions, that is, it has length, and allows the electrons to propagate in a particle-like fashion. It is, constructed typically on a semiconductor base.

SAM Self assembled monolayer. It is an organized layer of amphiphilic molecules in which one end of the molecule, the "head group", shows a special affinity for a substrate. SAMs also consist of a tail with a functional group at the terminal end. The hydrophilic "head groups" assemble together on the substrate, while the hydrophobic tail groups assemble far from the substrate. Areas of close-packed molecules nucleate and grow until the surface of the substrate is covered in a single monolayer.

Sol-gel process A wet-chemical technique (chemical solution deposition) widely used recently in the fields of materials science and ceramic engineering. Such methods are used primarily for the fabrication of materials (typically a metal oxide) starting from a chemical solution which acts as the precursor for an integrated network of either discrete particles or network polymers.

STM Scanning tunnelling microscopy. It a powerful technique for viewing surfaces at the atomic level. Its development in 1981 earned its inventors, Gerd Binnig and Heinrich Rohrer, the Nobel Prize in Physics in 1986. For STM, good resolution is considered to be 0.1 nm lateral resolution and 0.01 nm depth resolution. With this resolution, the individual atoms within materials are routinely imaged and manipulated.

Surfactant The term surfactant is a blend of surface active agent and it was coined by Antara in 1950. Surfactants are usually organic compounds that are amphiphilic, meaning they contain both hydrophobic groups (their "tails") and hydrophilic groups (their "heads"). Therefore, they are soluble in both organic solvents and water.

SWNT Single-walled nanotubes. They have a diameter close to 1 nanometre, with a tube length that can be many millions of times longer. The structure of an SWNT can be conceptualized by wrapping a one-atom-thick layer of graphite called graphene into a seamless cylinder.

Top-down approach Breaking down a system to gain insight into its compositional sub-systems. In a top-down approach an overview of the system is first formulated, specifying but not detailing any first-level subsystems. Each subsystem is then refined in yet greater detail, sometimes in many additional subsystem levels, until the entire specification is reduced to base elements.

Wet Nanotechnology The study of biological systems that exist primarily in a water environment. The functional nanometer-scale structures of interest here are genetic material, membranes, enzymes and other cellular components. The success of this nanotechnology is amply demonstrated by the existence of living organisms whose form, and evolution are governed by the interactions of nanometre-scale structures.

INDEX

www.ingramcontent.com/pod-product-compliance
Lightning Source LLC
Chambersburg PA
CBHW081126180926
51611CB00005B/69
9788180940200